SpringerBriefs on Cyber Security Systems and Networks

Editor-in-chief

Yang Xiang, Digital Research & Innovation Capability Platform, Swinburne University of Technology, Hawthorn, Melbourne, VIC, Australia

Series editors

Liqun Chen ⓘ, University of Surrey, Guildford, UK
Kim-Kwang Raymond Choo, Department of Information Systems and Cyber Security, University of Texas at San Antonio, San Antonio, TX, USA
Sherman S. M. Chow, Department of Information Engineering, The Chinese University of Hong Kong, Hong Kong
Robert H. Deng, School of Information Systems, Singapore Management University, Singapore, Singapore
Dieter Gollmann, Hamburg University of Technology, Hamburg, Germany
Javier Lopez, University of Malaga, Malaga, Spain
Kui Ren, University at Buffalo, Buffalo, NY, USA
Jianying Zhou, Singapore University of Technology and Design, Singapore, Singapore

The series aims to develop and disseminate an understanding of innovations, paradigms, techniques, and technologies in the contexts of cyber security systems and networks related research and studies. It publishes thorough and cohesive overviews of state-of-the-art topics in cyber security, as well as sophisticated techniques, original research presentations and in-depth case studies in cyber systems and networks. The series also provides a single point of coverage of advanced and timely emerging topics as well as a forum for core concepts that may not have reached a level of maturity to warrant a comprehensive textbook. It addresses security, privacy, availability, and dependability issues for cyber systems and networks, and welcomes emerging technologies, such as artificial intelligence, cloud computing, cyber physical systems, and big data analytics related to cyber security research. The mainly focuses on the following research topics:

Fundamentals and Theories

- Cryptography for cyber security
- Theories of cyber security
- Provable security

Cyber Systems and Networks

- Cyber systems Security
- Network security
- Security services
- Social networks security and privacy
- Cyber attacks and defense
- Data-driven cyber security
- Trusted computing and systems

Applications and Others

- Hardware and device security
- Cyber application security
- Human and social aspects of cyber security

More information about this series at http://www.springer.com/series/15797

Kwangjo Kim • Muhamad Erza Aminanto
Harry Chandra Tanuwidjaja

Network Intrusion Detection using Deep Learning

A Feature Learning Approach

 Springer

Kwangjo Kim
School of Computing (SoC)
Korea Advanced Institute of
Science and Technology
Daejeon, Korea (Republic of)

Muhamad Erza Aminanto
School of Computing (SoC)
Korea Advanced Institute of
Science and Technology
Daejeon, Korea (Republic of)

Harry Chandra Tanuwidjaja
School of Computing (SoC)
Korea Advanced Institute of
Science and Technology
Daejeon, Korea (Republic of)

ISSN 2522-5561   ISSN 2522-557X   (electronic)
SpringerBriefs on Cyber Security Systems and Networks
ISBN 978-981-13-1443-8   ISBN 978-981-13-1444-5   (eBook)
https://doi.org/10.1007/978-981-13-1444-5

Library of Congress Control Number: 2018953758

This Springer imprint is published by the registered company Springer Nature Singapore Pte Ltd.
The registered company address is: 152 Beach Road, #21-01/04 Gateway East, Singapore 189721, Singapore

To our families for their lovely support.

Preface

This monograph presents recent advances in Intrusion Detection System (IDS) using deep learning models, which have achieved great success recently, particularly in the field of computer vision, natural language processing, and image processing. The monograph provides a systematic and methodical overview of the latest developments in deep learning and makes a comparison among deep learning-based IDSs. A comprehensive overview of deep learning applications to IDS followed by deep feature learning methods containing a novel deep feature extraction and selection and deep learning for clustering is provided in this monograph. Further challenges and research directions are delivered in the monograph.

The monograph offers a rich overview of deep learning-based IDS, which is suitable for students, researchers, and practitioners interested in deep learning and intrusion detection and as a reference book. The comprehensive comparison of various deep-learning applications helps readers with a basic understanding of machine learning and inspires applications in IDS and other cybersecurity areas.

The outline of this monograph is as follows:

Chapter 1 describes the importance of IDS in computer networks these days by providing a survey of a security breach in computer networks. It is highlighted that deep learning models can improve IDS performance. It also explains the motivation of surveying deep learning-based IDSs.

Chapter 2 provides all the relevant definition of IDS. It then explains different types of the current IDS, based on where we put the detection module and based on the used approach. Common performance metrics and publicly available benchmark dataset are also provided in this chapter.

Chapter 3 provides a brief preliminary study regarding classical machine learning which consists of supervised, unsupervised, semi-supervised, weakly supervised, reinforcement, and adversarial machine learning. It briefly surveys 22 papers which are using machine learning techniques for their IDSs.

Chapter 4 discusses several deep learning models which contain generative, discriminative, and hybrid approaches.

Chapter 5 surveys various IDSs that leverage deep learning models which are divided into four classes: generative, discriminative, hybrid, and deep reinforcement learning.

Chapter 6 discusses the importance of deep learning models as a feature learning (FL) approach in IDS researches. We explain further two models which are deep feature extraction and selection and deep learning for clustering.

Chapter 7 concludes this monograph by providing an overview of challenges and future research directions in deep learning applications for IDS.

Appendix discusses several papers of malware detection over a network using deep learning models. Malware detection is also an important issue due to the increasing number of malware and similar approach as IDS.

Daejeon, Republic of Korea   Kwangjo Kim
March 2018   Muhamad Erza Aminanto
                                                            Harry Chandra Tanuwidjaja

Acknowledgments

This monograph was partially supported by the Institute for Information & Communications Technology Promotion (IITP) grant funded by the Korea Government (MSIT) (2013-0-00396, Research on Communication Technology using Bio-Inspired Algorithm, and 2017-0-00555, Towards Provable-secure Multi-party Authenticated Key Exchange Protocol based on Lattices in a Quantum World) and by the National Research Foundation of Korea (NRF) grant funded by the Korea Government (MSIT) (No. NRF-2015R1A-2A2A01006812).

We are very grateful for Prof. Hong-Shik Park, School of Electrical Engineering, KAIST, who gave us the excellent opportunity to execute this research initiative by combining deep learning into intrusion detection for secure wireless network. We also thank Prof. Paul D. Yoo and Prof. Taufiq Asyhari, Cranfield Defence and Security, UK, who gave us inspiring discussion during our research working.

The authors sincerely appreciate the contribution of the alumni and current member of the Cryptology and Information Security Lab. (CAISLAB), Graduate School of Information Security, School of Computing, KAIST, Khalid Huseynov, Dongsoo Lee, Kyungmin Kim, Hakju Kim, Rakyong Choi, Jeeun Lee, Soohyun Ahn, Joonjeong Park, Jeseoung Jung, Jina Hong, Sungsook Kim, Edwin Ayisi Opare, Hyeongcheol An, Seongho Han, Nakjun Choi, Nabi Lee, and Dongyeon Hong.

We gratefully acknowledge the editors of this monograph series on security for their valuable comments and the Springer to give us to write this monograph.

Finally we are also very grateful for our families for their strong support and endless love.

Contents

Acronyms

ACA	Ant Clustering Algorithm
ACC	Ant Colony Clustering
AE	Auto-Encoder
AIS	Artificial Immune System
ANN	Artificial Neural Network
APT	Advanced Persistent Threat
ATTA-C	Adaptive Time-Dependent Transporter Ants Clustering
AWID	Aegean Wi-Fi Intrusion Dataset
BM	Boltzmann Machine
CAN	Controller Area Network
CCN	Content-Centric Network
CFS	CfsSubsetEval
CNN	Convolutional Neural Network
CoG	Center of Gravity
Corr	Correlation
CPS	Cyber-Physical System
DAE	Denoising Auto-Encoder
DBM	Deep Boltzmann Machine
DBN	Deep Belief Network
DDoS	Distributed Denial of Service
D-FES	Deep Feature Extraction and Selection
DNN	Deep Neural Network
DoS	Denial of Service
DR	Detection Rate
DT	Decision Tree
ERL	Evolutionary Reinforcement Learning
ESVDF	Enhanced Support Vector Decision Function
FIS	Fuzzy Inference System
FL	Feature Learning
FN	False Negative
FNR	False Negative Rate

FP	False Positive
FPR	False Positive Rate
FW	Firewall
GAN	Generative Adversarial Networks
GPU	Graphics Processing Unit
GRU	Gated Recurrent Unit
HJI	Hamiltonian-Jacobi-Isaac
HIS	Human Inference System
ICV	Integrity Check Value
IDS	Intrusion Detection System
IG	Information Gain
IoT	Internet of Things
IPS	Intrusion Prevention System
IV	Initialization Vector
JSON	Java Script Object Notation
KL	Kullback-Leibler
kNN	K-Nearest Neighbors
LoM	Largest of Max
LSTM	Long Short-Term Memory
MDP	Markov Decision Processes
METIS	Mobile and Wireless Communications Enablers for the Twenty-Twenty Information Society
MF	Membership Functions
MLP	Multi-Layer Perceptron
MoM	Mean of Max
MSE	Mean Square Error
NN	Neural Network
PSO	Particle Swarm Optimization
R2L	Remote to Local
RBM	Restricted Boltzmann Machine
RL	Reinforcement Learning
RNN	Recurrent Neural Networks
SAE	Stacked Auto-Encoder
SDAE	Stacked Denoising Auto-Encoder
SDN	Software-Defined Networking
SFL	Supervised Feature Learning
SGD	Stochastic Gradient Descent
SNN	Shared Nearest Neighbor
SOM	Self-Organizing Map
SoM	Smallest of Max
SPN	Sum-Product Networks
STL	Self-Taught Learning
SVM	Support Vector Machine
SVM-RFE	SVM-Recursive Feature Elimination
TBM	Time to Build Model

TCP/IP	Transmission Control Protocol/Internet Protocol
TN	True Negative
TP	True Positive
TT	Time to Test
U2R	User to Root
UFL	Unsupervised Feature Learning

Chapter 1
Introduction

Abstract This chapter discusses the importance of IDS in computer networks while wireless networks grow rapidly these days by providing a survey of a security breach in wireless networks. Many methods have been used to improve IDS performance, the most promising one is to deploy machine learning. Then, the usefulness of recent models of machine learning, called a deep learning, is highlighted to improve IDS performance, particularly as a Feature Learning (FL) approach. We also explain the motivation of surveying deep learning-based IDSs.

Computer networks and Internet are inseparable from human life today. Abundant applications rely on Internet, including life-critical applications in healthcare and military. Moreover, extravagant financial transactions exist over the Internet every day. This rapid growth of the Internet has led to a significant increase in wireless network traffic in recent years. According to a worldwide telecommunication consortium, Mobile and Wireless Communications Enablers for the Twenty-Twenty Information Society (METIS) [1], a proliferation of 5G and Wi-Fi networks is expected to occur in the next decades. They believe that avalanche of mobile and wireless traffic volume will occur due to the development of society needs to be fulfilled. Applications such as e-learning, e-banking, and e-health would spread and become more mobile. By 2020[1] wireless network traffic is anticipated to account for two-thirds of total Internet traffic—with 66% of IP traffic expected to be generated by Wi-Fi and cellular devices only.

Cyber-attacks have become an immense growing rate as Internet of Things (IoT) are widely used these days [2].

IBM [3] reported an enormous account hijacked during 2016, and spam emails are four times higher than the previous year. Common attacks noticed in the same

---

[1]Cisco Visual Networking Index: Forecast and Methodology 2015–2020, published at www.cisco.com/c/en/us/solutions/collateral/service-provider/visual-networking-index-vni/complete-white-paper-c11-481360.html

© The Author(s), under exclusive license to Springer Nature Singapore Pte Ltd. 2018   1
K. Kim et al., *Network Intrusion Detection using Deep Learning*,
SpringerBriefs on Cyber Security Systems and Networks,
https://doi.org/10.1007/978-981-13-1444-5_1

report include brute-force, malvertising, phishing, SQL injection, DDoS, malware, etc. Majority of malwares are accounted as ransomware (85% of malwares existed in a year are ransomware). These attacks might leak sensitive data or disrupt normal operations which lead to an enormous financial loss. The most popular companies impacted by security incidents are financial services-related companies, followed by information and communications, manufacture, retail, and healthcare [3]. Wireless networks such as IEEE 802.11 have been widely deployed to provide users with mobility and flexibility in the form of high-speed local area connectivity. However, other issues such as privacy and security have raised. The rapid spread of IoT-enabled devices has resulted in wireless networks becoming to both passive and active attacks, the number of which has grown dramatically [2]. Examples of these attacks are impersonation, flooding, and injection attacks. The wide and rapid spread of computing devices using Wi-Fi networks creates complex, large, and high-dimensional data, which cause confusion when capturing attack properties and force us to strengthen our security measures in our system.

Comprehensive researches have been executed to avoid attacks as mentioned earlier. An Intrusion Detection System (IDS) is one of the most common components of every network security infrastructure [4] including wireless networks [5]. Machine-learning techniques have been well adopted as the primary detection algorithm in IDS owing to their model-free properties and learnability [6]. Leveraging the recent development of machine-learning techniques such as deep learning [7] can be expected to bring significant benefits on improving existing IDSs particularly for detecting impersonation attacks in large-scale networks. Based on the detection method, IDS can be classified into three types: misuse-, anomaly-, and specification-based IDS. A misuse-based IDS also known as a signature-based IDS [8] detects any attack by checking whether the attack characteristics match previously stored signatures or patterns of attacks. This type of IDS is suitable for detecting known attacks; however, new or unknown attacks are difficult to detect.

Extensive research applying machine-learning methods in IDS has been done in both academia and industry. However, the security experts are still pursuing better performance IDS which has the highest Detection Rate (DR) and the lowest false alarm rate. Also, overall threat analysis is expected to secure their networks [9]. Improvements in IDS could be achieved by embracing a recent breakthrough in machine learning [6], so-called deep learning. Deep Learning applications have won numerous contests in pattern recognition and machine learning [10]. Deep learning belongs to a class of machine-learning methods, which employs consecutive layers of information-processing stages in hierarchical manners for pattern classification and feature or representation learning [11]. According to [12], there are three important reasons for the deep learning prominence recently. First, processing abilities (e.g., GPU units) has been increased sharply. Second, computing hardware is getting affordable, and the third is a recent breakthrough in machine-learning research. Shallow and deep learners are distinguished by the depth of their credit assignment paths, which are chains of possibly learnable, causal links between actions and effects. Usually, deep learning plays an important role in image classification results. Besides, deep learning is also commonly used for language, graphical modeling, pattern recognition, speech, audio, image, video, natural language, and signal processing [11]. There are many deep learning methods such as Deep Belief Network (DBN), Boltzmann Machine (BM), Restricted Boltzmann Machine (RBM), Deep Boltzmann Machine (DBM), Deep Neural Network (DNN), auto-encoder, Deep Auto-Encoder (DAE), Stacked Auto-Encoder (SAE), Stacked Denoising Auto-Encoder (SDAE), distributed representation, and Convolutional Neural Network (CNN). One of the distinguished applications is AlphaGo [13] from Google that uses CNN. AlphaGo beat the Korean world champion in the "Go" game recently by showing superman-like capabilities in remote machine learning. The advancements in learning algorithms might improve the performance of IDS to reach higher DR and lower false alarm rate.

The broad and rapid spread of computing devices using the Internet, especially Wi-Fi networks, creates complex, large, and high-dimensional data, which cause inevitable confusions when capturing attack properties. FL acts as an essential tool for improving the learning process of a machine-learning model. It consists of feature construction, extraction, and selection. Feature construction expands the original features to enhance their expressiveness, whereas feature extraction transforms the original features into a new form, and feature selection eliminates unnecessary features [14]. FL is a key to improve the performance of existing machine-learning-based IDSs.

We realize that there is a confusion of how to adopt deep learning in IDS applications properly since the different approaches have adopted by each previous one. Several researches use deep learning methods in a partial sense only, while the rest still uses conventional neural networks. The complexity of deep learning method may be one of the reasons. Also, deep learning method requires a lot of time to train correctly. However, we found that several researchers adopt deep learning method for feature learning and classification for an intelligent IDS in their network. We compare the IDS performance among them.

Tran et al. [15] provided one example on how to use deep learning in IDS. Classical machine-learning algorithms, Naive Bayes and C4.5, assisted by high-level features were generated using genetic programming. This approach is a common way of leveraging deep learning models in IDS, where the deep learning models are assisting any classical machine learning with high-level features. This approach is also adopted by Aminanto et al. [16] which is explained further in Chap. 6. In this monograph, we highlighted some IDSs which use deep learning models. Hamed et al. [17] surveyed several preprocessing techniques in IDS researches, how to collect data from real world and honeypot and how to build a dataset from raw input data. Although most of IDSs use deep learning models as their data preprocessing technique, this monograph focuses on reviewing deep learning-based IDSs.

Deep learning is beneficial in IDS, especially for FL. This monograph examines such deep learning methods with their pros and cons to get a better understanding of how to apply deep learning in IDS. At the end, we provide future challenges and directions to employ deep learning in IDS accordingly.

References

1. A. Osseiran, F. Boccardi, V. Braun, K. Kusume, P. Marsch, M. Maternia, O. Queseth, M. Schellmann, H. Schotten, H. Taoka, H. Tullberg, M. A. Uusitalo, B. Timus, and M. Fallgren, "Scenarios for 5G mobile and wireless communications: The vision of the metis project," *IEEE Commun. Mag.*, vol. 52, no. 5, pp. 26–35, May 2014.
2. C. Kolias, A. Stavrou, J. Voas, I. Bojanova, and R. Kuhn, "Learning internet-of-things security" hands-on"," *IEEE Security Privacy*, vol. 14, no. 1, pp. 37–46, 2016.
3. M. Alvarez, N. Bradley, P. Cobb, S. Craig, R. Iffert, L. Kessem, J. Kravitz, D. McMilen, and S. Moore, "IBM X-force threat intelligence index 2017," *IBM Corporation*, pp. 1–30, 2017.
4. C. Kolias, G. Kambourakis, and M. Maragoudakis, "Swarm intelligence in intrusion detection: A survey," *Computers & Security*, vol. 30, no. 8, pp. 625–642, 2011.
5. A. G. Fragkiadakis, V. A. Siris, N. E. Petroulakis, and A. P. Traganitis, "Anomaly-based intrusion detection of jamming attacks, local versus collaborative detection," *Wireless Communications and Mobile Computing*, vol. 15, no. 2, pp. 276–294, 2015.
6. R. Sommer and V. Paxson, "Outside the closed world: On using machine learning for network intrusion detection," in *Proc. Symp. Security and Privacy, Berkeley, California*. IEEE, 2010, pp. 305–316.
7. G. Anthes, "Deep learning comes of age," *Communications of the ACM*, vol. 56, no. 6, pp. 13–15, 2013.
8. A. H. Farooqi and F. A. Khan, "Intrusion detection systems for wireless sensor networks: A survey," in *Proc. Future Generation Information Technology Conference, Jeju Island, Korea*. Springer, 2009, pp. 234–241.
9. R. Zuech, T. M. Khoshgoftaar, and R. Wald, "Intrusion detection and big heterogeneous data: a survey," *Journal of Big Data*, vol. 2, no. 1, p. 3, 2015.
10. J. Schmidhuber, "Deep learning in neural networks: An overview," *Neural Networks*, vol. 61, pp. 85–117, 2015.
11. L. Deng, "A tutorial survey of architectures, algorithms, and applications for deep learning," *APSIPA Transactions on Signal and Information Processing*, vol. 3, 2014.
12. L. Deng, D. Yu, *et al.*, "Deep learning: methods and applications," *Foundations and Trends® in Signal Processing*, vol. 7, no. 3–4, pp. 197–387, 2014.
13. D. Silver, A. Huang, C. J. Maddison, A. Guez, L. Sifre, G. Van Den Driessche, J. Schrittwieser, I. Antonoglou, V. Panneershelvam, M. Lanctot, *et al.*, "Mastering the game of go with deep neural networks and tree search," *Nature*, vol. 529, no. 7587, pp. 484–489, 2016.
14. H. Motoda and H. Liu, "Feature selection, extraction and construction," *Communication of IICM (Institute of Information and Computing Machinery), Taiwan*, vol. 5, pp. 67–72, 2002.
15. B. Tran, S. Picek, and B. Xue, "Automatic feature construction for network intrusion detection," in *Asia-Pacific Conference on Simulated Evolution and Learning*. Springer, 2017, pp. 569–580.
16. M. E. Aminanto, R. Choi, H. C. Tanuwidjaja, P. D. Yoo, and K. Kim, "Deep abstraction and weighted feature selection for Wi-Fi impersonation detection," *IEEE Transactions on Information Forensics and Security*, vol. 13, no. 3, pp. 621–636, 2018.
17. T. Hamed, J. B. Ernst, and S. C. Kremer, "A survey and taxonomy on data and pre-processing techniques of intrusion detection systems," in *Computer and Network Security Essentials*. Springer, 2018, pp. 113–134.

Chapter 2
Intrusion Detection Systems

Abstract This chapter briefly introduces all the relevant definitions on Intrusion Detection System (IDS), followed by a classification of current IDSs, based on the detection module located and the approach adopted. We also explain and provide examples of one common IDS in research fields, which is machine-learning-based IDS. Then, we discuss an example of IDS using bio-inspired clustering method.

2.1 Definition

An IDS becomes a standard security measure in computer networks. Unlike Firewall (FW) in Fig. 2.1a, IDS is usually located inside the network to monitor all internal traffics as shown in Fig. 2.1b. One may consider using both FW and IDS to protect the network efficiently. IDS is defined autonomous process of intrusion detection which is to find events of violation of security policies or standard security practices in computer networks [1]. Besides identifying the security incidents, IDS also has other functions: documenting existing threats and deterring adversaries [1]. IDS requires particular properties which acts as a passive countermeasure, monitors whole or part of networks only, and aims high attack detection rate and low false alarm rate.

2.2 Classification

We can divide IDSs depending on the placement and the methodology deployed in a network. By the positioning of the IDS module in the network, we might distinguish IDSs into three classes: network-based, host-based, and hybrid IDSs. The first IDS, network-based IDS as shown in Fig. 2.2, puts the IDS module inside the network where whole can be monitored. This IDS checks for malicious activities by inspecting all packets moving across the network. On the other hand, Fig. 2.3 shows

K. Kim et al., *Network Intrusion Detection using Deep Learning*,
SpringerBriefs on Cyber Security Systems and Networks,
https://doi.org/10.1007/978-981-13-1444-5_2

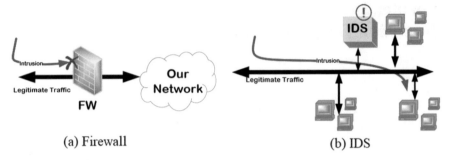

(a) Firewall   (b) IDS

Fig. 2.1 Typical network using (**a**) Firewall and (**b**) IDS

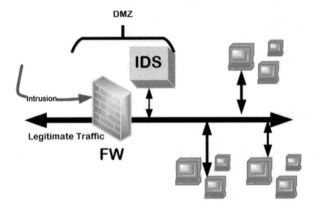

Fig. 2.2 Network-based IDS

the host-based IDS which places the IDS module on each client of the network. The module examining all inbound and outbound traffics of the corresponding client leads to detailed monitoring of the particular client. Two types of IDSs have specific drawbacks—the network-based IDS might burden the workload and then miss some malicious activities, while the host-based IDS does not monitor all the network traffics, having less workload than the network-based IDS. Therefore, the hybrid IDS as shown in Fig. 2.4 places IDS modules in the network as well as clients to monitor both specific clients and network activities at the same time.

Based on the detection method, IDSs can be divided into three different types: misuse-, anomaly-, and specification-based IDSs. A misuse-based IDS, known as a signature-based IDS [2], looks for any malicious activities by matching the known signatures or patterns of attacks with the monitored traffics. This IDS suits for known attack detection; however, new or unknown attacks (also called as a zero-day exploit) are difficult to be detected. An anomaly-based IDS detects an attack by profiling normal behavior and then triggers an alarm if there is any deviation from it. The strength of this IDS is its capability for unknown attack detection. Misuse-based IDS usually achieves higher detection performance for known attacks than anomaly-based IDS. A specification-based IDS manually defines a set of rules and constraints to express the normal operations. Any deviation from the rules and the

Fig. 2.3 Host-based IDS

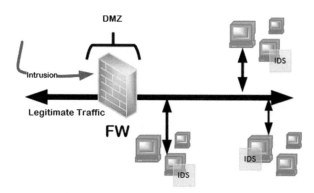

Fig. 2.4 Hybrid IDS

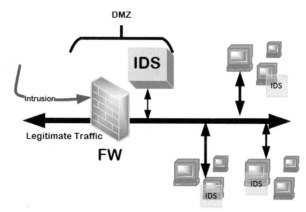

Table 2.1 Comparison of IDS types based on the methodology

	Misuse-based	Anomaly-based	Specification-based
Method	Identify known attack patterns	Identify unusual activity patterns	Identify violation of predefined rules
DR	High	Low	High
False alarm rate	Low	High	Low
Unknown attack detection	Incapable	Capable	Incapable
Drawback	Updating signatures is burdensome	Computing any machine learning is heavy	Relying on expert knowledge during defining rules is undesirable

constraints during the execution is flagged as malicious [3]. Table 2.1 summarizes the comparison of IDS types based on the methodology.

We discuss further machine-learning-based IDS which belongs to the anomaly-based IDS [4]. There are two types of learning, namely, supervised and unsupervised learning. The unsupervised learning does not require a labeled dataset for training which is crucial for colossal network traffics recently, while the supervised learning requires a labeled dataset. Unsupervised learning capability is of critical significance as it allows a model to detect new attacks without creating costly labels or dependent

Table 2.2 Comparison between supervised and unsupervised learning

	Supervised	Unsupervised
Definition	The dataset are labeled with predefined classes	The dataset are labeled **without** predefined classes
Approach	Classification	Clustering
Method	Support Vector Machine, Decision Tree, etc.	K-means clustering, Ant Clustering Algorithm etc.
Known attack detection	High	Low
Unknown attack detection	Low	High

variables. Table 2.2 outlines the comparison between supervised and unsupervised learning.

2.3   Benchmark

This section discusses benchmarking techniques in IDS researches. The benchmark dataset and the uniform performance metrics are essential to evaluate and compare two or more models. By the fair comparison, we can identify the improvement made by any proposed approach.

2.3.1   Performance Metric

The evaluation of any IDS performance can be done by adopting the common model performance measures [5]: accuracy (Acc), DR, false alarm rate (FAR), Mcc, $Precision$, F_1 score, CPU Time to Build Model (TBM), and CPU Time to Test (TT). Acc shows the overall effectiveness of an algorithm [6]. DR, also known as $Recall$, refers to the number of impersonation attacks detected divided by the total number of impersonation attack instances in the test dataset. Unlike $Recall$, $Precision$ counts the number of impersonation attacks detected among the total number of instances classified as an attack. The F_1 score measures the harmonic mean of $Precision$ and $Recall$. FAR is the number of normal instances classified as an attack divided by the total number of normal instances in the test dataset, while FNR shows the number of attack instances that are unable to be detected. Mcc represents the correlation coefficient between the detected and observed data [7]. Intuitively, the goal is to achieve a high Acc, DR, $Precision$, Mcc, and F_1 score and, at the same time, maintain low FAR, TBM, and TT. The above measures can be defined by Eqs. (2.1), (2.2), (2.3), (2.4), (2.5), (2.6), and (2.7):

$$Acc = \frac{TP + TN}{TP + TN + FP + FN}, \tag{2.1}$$

$$DR(Recall) = \frac{TP}{TP + FN}, \tag{2.2}$$

$$Precision = \frac{TP}{TP + FP}, \tag{2.3}$$

$$FAR = \frac{FP}{TN + FP}, \tag{2.4}$$

$$FNR = \frac{FN}{FN + TP}, \tag{2.5}$$

$$F_1 = \frac{2TP}{2TP + FP + FN}, \tag{2.6}$$

$$Mcc = \frac{(TP \times TN) - (FP \times FN)}{\sqrt{(TP + FP)(TP + FN)(TN + FP)(TN + FN)}}, \tag{2.7}$$

where True Positive (TP) is the number of intrusions correctly classified as an attack, True Negative (TN) is the number of normal instances correctly classified as a benign packet, False Negative (FN) is the number of intrusions incorrectly classified as a benign packet, and False Positive (FP) is the number of normal instances incorrectly classified as an attack.

2.3.2   Public Dataset

The benchmarking dataset is one important aspect to evaluate the effectiveness of a proposed model of IDS. KDD Cup'99 dataset has been the most popular dataset for the evaluation of anomaly detection methods [8]. This dataset is based on the data captured in DARPA'98 IDS evaluation program, which consists of approximately 4,900,000 single connection instances. Table 2.3 shows the packet distribution of KDD Cup 99 dataset [9]. Each instance contains 41 features and is labeled as either normal or attack instance. The dataset provides four distinct attack types as follows:

1. **Probing Attack**: An attacker attempts to collect information about computer networks to bypass the security controls. An example of probing attack is port scanning.
2. **Denial of Service (DoS) Attack**: An attack in which the attacker prevents legitimate users from accessing authorized data. The attacker aims to compute resources too exhausted to handle legitimate requests by flooding the network with unnecessary packet requests. An example of DoS attack is SYN flood attack.
3. **User to Root (U2R) Attack**: An attacker starts the attack by accessing to a normal user account on the system. Then, the attacker exploits the vulnerability to gain root access to the system. An example of U2R attack is *xterm* exploitation.

Table 2.3 Packet
distribution of KDD Cup'99
dataset

Type	# of Packets	Proportion (%)
Normal	972,781	19.86
Probe	41,102	0.84
DoS	3,883,370	79.28
U2R	52	0.00
R2L	1,126	0.02
Total	4,898,431	100

4. **Remote to Local (R2L) Attack**: This kind of attack is executed by an attacker who is capable of sending packets to a machine over a network but does not have an account on that machine. The attacker exploits some vulnerabilities to gain local access as a user of that machine remotely. An example of R2L attack is ftp_write exploitation.

Despite the usefulness of KDD Cup'99 dataset, this dataset has some statistical drawbacks which are redundant instances and irrational distribution of particular class in both training and test datasets. Therefore, NSL-KDD dataset was developed to refine the limitation of original KDD Cup'99 dataset [10]. The NSL-KDD dataset also has 41 attributes same as the original one.

In Wi-Fi network area, there is a dataset so-called Aegean Wi-Fi Intrusion Dataset (AWID) developed by Kolias et al. [11]. There are two types of AWID dataset. The first type, named "CLS," has four target classes, whereas the second, named "ATK," has 16 target classes. The 16 classes of the "ATK" dataset belong to the four attack categories in the "CLS" dataset. As an example, the *Caffe-Latte*, *Hirte*, *Honeypot*, and *EvilTwin* attack types, listed in the "ATK" dataset, are categorized as an impersonation attack in the "CLS" dataset. Based on the size of the data instances included, the AWID dataset comprises both full and reduced versions. There are 1,795,595 instances in the reduced training dataset, with 1,633,190 and 162,385 normal and attack instances, respectively. There are 575,643 instances in the reduced test dataset, with 530,785 and 44,858 normal and attack instances, respectively, as shown in Table 6.1 (see Chap. 6).

References

1. K. Scarfone and P. Mell, "Guide to intrusion detection and prevention systems (idps)," *NIST special publication*, vol. 800, no. 2007, 2007.
2. A. H. Farooqi and F. A. Khan, "Intrusion detection systems for wireless sensor networks: A survey," in *Proc. Future Generation Information Technology Conference, Jeju Island, Korea*. Springer, 2009, pp. 234–241.
3. R. Mitchell and I. R. Chen, "Behavior rule specification-based intrusion detection for safety critical medical cyber physical systems," *IEEE Trans. Dependable Secure Comput.*, vol. 12, no. 1, pp. 16–30, Jan 2015.
4. I. Butun, S. D. Morgera, and R. Sankar, "A survey of intrusion detection systems in wireless sensor networks," *IEEE Commun. Surveys Tuts.*, vol. 16, no. 1, pp. 266–282, 2014.

5. O. Y. Al-Jarrah, O. Alhussein, P. D. Yoo, S. Muhaidat, K. Taha, and K. Kim, "Data randomization and cluster-based partitioning for botnet intrusion detection," *IEEE Trans. Cybern.*, vol. 46, no. 8, pp. 1796–1806, 2015.

6. M. Sokolova, N. Japkowicz, and S. Szpakowicz, "Beyond accuracy, F-score and ROC: a family of discriminant measures for performance evaluation," in *Australasian Joint Conference on Artificial Intelligence, Hobart, Australia.* Springer, 2006, pp. 1015–1021.

7. P. D. Schloss and S. L. Westcott, "Assessing and improving methods used in operational taxonomic unit-based approaches for 16s rRNA gene sequence analysis," *Applied and environmental microbiology*, vol. 77, no. 10, pp. 3219–3226, 2011.

8. M. Tavallaee, E. Bagheri, W. Lu, and A.-A. Ghorbani, "A detailed analysis of the kdd cup 99 data set," *Proceedings of the Second IEEE Symposium on Computational Intelligence for Security and Defence Applications 2009*, pp. 53–58, 2009.

9. H. M. Shirazi, "An intelligent intrusion detection system using genetic algorithms and features selection," *Majlesi Journal of Electrical Engineering*, vol. 4, no. 1, 2010.

10. S. Potluri and C. Diedrich, "Accelerated deep neural networks for enhanced intrusion detection system," in *Emerging Technologies and Factory Automation (ETFA), 2016 IEEE 21st International Conference on.* IEEE, 2016, pp. 1–8.

11. C. Kolias, G. Kambourakis, A. Stavrou, and S. Gritzalis, "Intrusion detection in 802.11 networks: empirical evaluation of threats and a public dataset," *IEEE Commun. Surveys Tuts.*, vol. 18, no. 1, pp. 184–208, 2015.

Chapter 3
Classical Machine Learning and Its Applications to IDS

Abstract This chapter provides a brief preliminary study regarding classical machine learning which consists of six different models: supervised, unsupervised, semi-supervised, weakly supervised, reinforcement, and adversarial machine learning. Then, the 22 papers are surveyed, which use machine-learning techniques for their IDSs.

3.1 Classification of Machine Learning

Different types of machine-learning models were leveraged in anomaly-based IDS. This chapter provides a preliminary study regarding classical machine-learning models. Machine learning can be divided into five different models based on training data types: supervised, unsupervised, semi-supervised, weakly supervised, and Reinforcement Learning (RL). The following sub-chapters explain further for each model. Also, we discuss several machine-learning-based IDSs.

3.1.1 Supervised Learning

In supervised learning, target class data are necessary during training. The network would build a model based on the correctness of matching to the label data. There are many machine-learning models in supervised learning, some of them are as follows.

3.1.1.1 Support Vector Machine

A supervised SVM is usually used for classification or regression tasks. If n is the number of input features, the SVM plots each feature value as a coordinate point in

    13
K. Kim et al., *Network Intrusion Detection using Deep Learning*,
SpringerBriefs on Cyber Security Systems and Networks,
https://doi.org/10.1007/978-981-13-1444-5_3

n-dimensional space. Subsequently, a classification process is executed by finding the hyperplane that distinguishes two classes. Although SVM can handle a nonlinear decision border of arbitrary complexity, we use a linear SVM since the nature of the dataset can be investigated by linear discriminant classifiers. The decision boundary for linear SVM is a straight line in two-dimensional space. The main computational property of SVM is the support vectors which are the data points that lie closest to the decision boundary. The decision function, $D(\mathbf{x})$ [1], of the input vector $\mathbf{x}$ as expressed by Eq. (3.1) heavily depends on the support vectors.

$$D(\mathbf{x}) = \mathbf{w}\mathbf{x} + b \tag{3.1}$$

$$\mathbf{w} = \sum_k \alpha_k y_k \mathbf{x}_k \tag{3.2}$$

$$b = (y_k - \mathbf{w}\mathbf{x}_k), \tag{3.3}$$

where $\mathbf{w}$, y, α, and b denote the weight vector, the class label, the marginal support vector, and the bias value, respectively. k denotes the number of samples. Equations (3.2) and (3.3) show how to compute the value of $\mathbf{w}$ and b, respectively.

SVM-Recursive Feature Elimination (RFE) is an application of RFE using the magnitude of the weight to perform rank clustering [1]. The RFE ranks the feature set and eliminates the low-ranked features which contribute less than the other features for classification task [2].

3.1.1.2   Decision Tree

C4.5 is robust to noise data and able to learn disjunctive expressions [3]. It has a k-ary tree structure, which can represent a test of attributes from the input data by each node. Every branch of the tree shows potentially selected important features as the values of nodes and different test results. C4.5 uses a greedy algorithm to construct a tree in a top-down recursive divide-and-conquer approach [3]. The algorithm begins by selecting the attributes that yield the best classification result. This is followed by generating a test node for the corresponding attributes. The data are then divided based on the Information Gain (IG) value of the nodes according to the test attributes that reside in the parent node. The algorithm terminates when all data are grouped in the same class, or the process of adding additional separations produces a similar classification result, based on its predefined threshold.

3.1.2   Unsupervised Learning

Unlike supervised learning, unsupervised learning does not require any label data during training. This is one advantage of using unsupervised learning since building a comprehensive labeled dataset is usually tricky. Because of no target classes provided, the network looks for similar properties of each training instances and creates a group of those similar instances. For anomaly detection, one can consider the outlier instances as an anomaly. Some examples of unsupervised learning are as follows:

3.1.2.1   *K*-Means Clustering

K-means clustering algorithm groups all observations data into k clusters iteratively until the convergence will be reached. In the end, one cluster contains similar data since each data enters to the nearest cluster. K-means algorithm assigns a mean value of the cluster members as a cluster centroid. In every iteration, it calculates the shortest Euclidean distance from an observation data into any cluster centroid. Besides that, the intra-variances inside the cluster are also minimized by updating the cluster centroid iteratively. The algorithm would terminate when convergence is achieved, which the new clusters are the same as the previous iteration clusters [4].

3.1.2.2   Ant Clustering

Ant Clustering Algorithm (ACA) simulates random ant walks on a two-dimensional grid which objects are laid down at random [5]. Unlike the dimension of the input data, each data instance is randomly projected onto a cell of the grid. A grid cell can indicate the relative position of the data instance in the two-dimensional grid. The general idea of ACA is to keep similar items in their original N-dimensional space. Vizine et al. [5] assumed that each site or cell on the grid can be resided by at most one object, and one of the two following situations may occur: (i) one ant holds an object i and evaluates the probability of dropping it in its current position; (ii) an unloaded ant estimates the likelihood of picking up an object. An ant is selected randomly and can either pick up or drop an object at its current location [5].

The probability of picking up an object increases by disparity among objects in the surrounding area and vice versa. In contrast, the probability of dropping an object increases by high similarity among objects in the surrounding area. Vizine et al. [5] defined $d(i,j)$ in Eq. (3.4) as the Euclidean distance between objects i and j in their N-dimensional space. The density distribution function for object i, at a particular grid location, is defined by Eq. (3.4) as follows:

$$f(i) = \begin{cases} \frac{1}{s^2} \sum_j (1 - d(i, j)/\alpha) & f(i) > 0 \\ 0 & \text{Otherwise,} \end{cases} \qquad (3.4)$$

where s^2 is the number of cells in the surrounding area of i and α is a constant that depicts the disparity among objects. The $f(i)$ might reach maximum value when all the sites in the surrounding area are occupied by similar or even equal objects. The probability of picking up and dropping an object i is given by Eqs. (3.5) and (3.6), respectively:

$$P_{\text{pick}}(i) = \left(\frac{k_p}{k_p + f(i)}\right)^2, \qquad (3.5)$$

$$P_{\text{drop}}(i) = \begin{cases} 2f(i) & f(i) < k_d \\ 1 & \text{Otherwise,} \end{cases} \qquad (3.6)$$

where the parameters k_p and k_d are threshold constants of the probability of picking up and dropping an object, respectively. A loaded ant considers the first empty cell in its local area to drop the object, since the current position of the ant may be preoccupied by another object [5].

Tsang et al. [6] define two variables: intra-cluster and inter-cluster distance in order to measure ACA performance. High intra-cluster distance means better compactness. Meanwhile, large inter-cluster distance means better separateness. A good ACA should provide minimum intra-cluster distance and maximum inter-cluster distance to presents the inherent structures and knowledge from data patterns.

3.1.2.3   (Sparse) Auto-Encoder

An Auto-Encoder (AE) is a symmetric Neural Network (NN) model, which uses an unsupervised approach to build a model with non-labeled data, as shown in Fig. 3.1. AE extracts new features by using an encoder-decoder paradigm by running from inputs through the hidden layer only. This paradigm enhances its computational performance and validates that the code has captured the relevant information from the data. The encoder is a function that maps an input x to a hidden representation as expressed by Eq. (3.7).

$$y = s_f \left(W \cdot x + b_f \right), \qquad (3.7)$$

where s_f is a nonlinear activation function which is a decision-making function to determine the necessity of any feature. Mostly, a logistic sigmoid, $sig(t) = \dfrac{1}{1 + e^{-t}}$, is used as an activation function because of its continuity and differen-

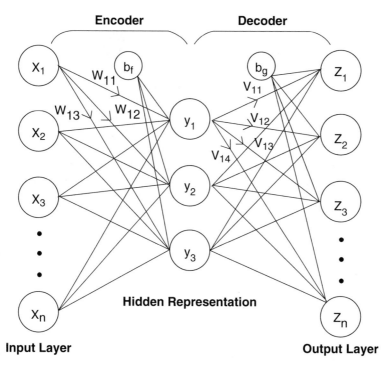

Fig. 3.1 AE network with symmetric input-output layers and three neurons in one hidden layer

tiability properties [7]. The decoder function expressed in Eq. (3.8) maps hidden representation y back to a reconstruction.

$$z = s_g \left(V \cdot y + b_g \right), \tag{3.8}$$

where s_g is the activation function of the decoder which commonly uses either the identity function, $s_g(t) = t$, or a sigmoid function such as an encoder. We use W and V acts as a weight matrix for the features. b_f and b_g acts as a bias vector for encoding and decoding, respectively. Its training phase finds optimal parameters $\theta = \{W, V, b_f, b_g\}$ which minimize the reconstruction error between the input data and its reconstruction output on a training set.

We further explain a modified form of AE, which is a sparse AE [8]. This is based on the experiments of Eskin et al. [9], in which anomalies usually form small clusters in scattered areas of feature space. Moreover, the dense and large clusters typically contain benign data [10]. For the sparsity of AE, we first observe the average output activation value of a neuron i, as expressed by Eq. (3.9).

$$\hat{\rho}_i = \frac{1}{N} \sum_{j=1}^{N} s_f \left(w_i^T x_j + b_{f,i} \right), \tag{3.9}$$

where N is the total number of training data, x_j is the j-th training data, w_i^T is the i-th row of a weight matrix W, and $b_{f,i}$ is the i-th row of a bias vector for encoding b_f. By lowering the value of $\hat{\rho}_i$, a neuron i in the hidden layer shows the specific feature presented in a smaller number of training data.

The task of machine learning is to fit a model to the given training data. However, the model often fits the particular training data but is incapable of classifying other data, and this is known as the overfitting problem. In this case, we can use a regularization technique to reduce the overfitting problem. The sparsity regularization Ω_{sparsity} evaluates how close the average output activation value $\hat{\rho}_i$ and the desired value ρ are, typically with *Kullback-Leibler* (KL) divergence to determine the difference between the two distributions, as expressed in Eq. (3.10).

$$
\begin{aligned}
\Omega_{\text{sparsity}} &= \sum_{i=1}^{h} KL(\rho \| \hat{\rho}_i) \\
&= \sum_{i=1}^{h} \left[\rho \log \left(\frac{\rho}{\hat{\rho}_i} \right) + (1 - \rho) \log \left(\frac{1 - \rho}{1 - \hat{\rho}_i} \right) \right],
\end{aligned}
\tag{3.10}
$$

where h is the number of neurons in the hidden layer.

We may increase the value of the entries of the weight matrix W to reduce the value of sparsity regularization. To avoid this situation, we also add regularization for the weight matrix, known as L_2 regularization as stated in Eq. (3.11).

$$
\Omega_{\text{weights}} = \frac{1}{2} \sum_{i=1}^{h} \sum_{j=1}^{N} \sum_{k=1}^{K} \left(w_{ji} \right)^2,
\tag{3.11}
$$

where N and K are the number of training data and the number of variables for each data, respectively.

The goal of training sparse AE is to find the optimal parameters, $\theta = \{W, V, b_f, b_g\}$, to minimize the cost function shown in Eq. (3.12).

$$
E = \frac{1}{N} \sum_{n=1}^{N} \sum_{k=1}^{K} (z_{kn} - x_{kn})^2 + \lambda \cdot \Omega_{\text{weights}} + \beta \cdot \Omega_{\text{sparsity}},
\tag{3.12}
$$

which is a regulated Mean Square Error (MSE) with L_2 regularization and sparsity regularization. The coefficient of the L_2 regularization term λ and the coefficient of sparsity regularization term β are specified while training the AE.

3.1.3 Semi-supervised Learning

Semi-supervised learning is a technique that can be illustrated "in the middle between supervised learning and unsupervised learning." The background of using semi-supervised learning in machine learning is because many researchers found that combining unlabeled data and a small amount of labeled data can improve the learning accuracy. We know that unlabeled data is cheap. On the other hand, labeled data can be hard to get, require experts with special devices, or consume too much time to process. So, we can say that the goal of semi-supervised learning is using both labeled and unlabeled data to build better learners, compared to using them separately. Since semi-supervised learning utilizes unlabeled data, an assumption must be made to generalize it. There are three kinds of assumptions [11]:

1. **Semi-supervised Smoothness Assumption**
   In this assumption, we consider the density of the input. If two points x and y are close from each other in a high-density region, so the corresponding output x' and y' are also in the same condition.
2. **Cluster Assumption**
   In this assumption, we consider the cluster of each input. If two points are in the same cluster, then they are likely to be in the same class.
3. **Manifold Assumption**
   In this assumption, we consider the dimension of the input. It says that high-dimensional data lie on a low-dimensional manifold.

There are three methods of semi-supervised learning [11]:

1. **Generative Models**
   When using generative models, we involve the estimation of conditional density. The distribution of data belongs to each class. As a result, the knowledge of problem's structure can be naturally integrated by modeling it.
2. **Low-Density Separation**
   When using low-density separation, we try to implement it by pushing the decision boundary away from the unlabeled points. We need to utilize maximum margin algorithm like Support Vector Machine (SVM) in this method. By using low-density separation, we can minimize the entropy.
3. **Graph Base Method**
   When using graph base method, we represent data by using nodes of a graph. We label the edge of each node with the pairwise distances of incident nodes. If there is any missing edge, it means that the edge corresponds to an infinite distance.

3.1.4 Weakly Supervised Learning

Weakly supervised learning is a machine-learning framework where the model is trained using examples which are partially annotated or labeled. Most modern computer vision system involves models learned from human-labeled image examples. As a result, we need to reduce the amount of human intervention required to train the models. Because of that reason, weakly supervised learning models try to make use of examples that are only partially labeled.

There are three types of weakly supervised learning [12]:

1. **Incomplete Supervision**

   In this method, only a subset (usually small) of training data is given with labels, while the other data remains unlabeled.
2. **Inexact Supervision**

   In this method, only coarse-grained labels are given. For example, let's consider an image categorization task. In the image, we will do image-level label rather than an object-level label.
3. **Inaccurate Supervision**

   In this method, the given labels are not always ground truth. It can happen when an image annotator is careless or some images cannot be categorized due to its ambiguity.

3.1.5 Reinforcement Learning

RL adopts a different approach to supervised and unsupervised learning. In RL, an agent was assigned to take the responsibility of target classes in supervised learning which provides the correct corresponding class for each instance. The agent is responsible for deciding which action to perform its task [13]. Since no training data involved, the agent learns from experience during RL training. The learning process adopts trial and error approach during the training to achieve a goal which is to earn a long-term and highest reward. RL training often illustrated as a way for the player to win a game with some small points along the way and an ultimate prize at the end of the way. The player (or the RL agent) will explore ways to reach the final award. Sometimes, the agent will be stuck in a small point due to exploiting the minor points as its goal. Therefore, in RL, the exploration of new ways is needed although small points are reached. By this, the agent can achieve the ultimate prize at the end. In other words, the agent and the game environment form a cyclic path of information which the agent does action, and the environment provides feedback based on the corresponding action. The process of reaching the ultimate prize can be formalized using Markov Decision Processes (MDP), where we can model the transition probabilities for each state. The objective function now is well represented by using MDP. There are two standard solutions to achieve this objective functions, Q-learning and policy learning. The former learns based on action-value function, while the latter learns using policy function which is a mapping between the best action and the corresponding state. More detail explanations are well delivered in [13].

3.1.6 Adversarial Machine Learning

This type of machine learning is different with all previous sorts, where the adversarial machine learning exploits and attacks the vulnerability of existing machine-learning models. One example was mentioned by Laskov and Lippman [14] which the capability foundations of machine learning are based on the assumption of the expressiveness of training data addressed by learning. However, this assumption might be violated if either the training or the test data distribution is modified intentionally to confuse the learning. Other examples of attacks occurred in an adversarial framework are also provided in [14].

In general, two threat models become the machine-learning risks [15]. The first threat is evasion attacks which try to bypass the learning result. The adversary attempts to evade from pattern matching done by machine-learning models. Spam filtering and intrusion detection are examples when the adversary wants to evade. Their basic idea to accomplish evasion attack is a trial-and-error approach until a given instance successfully evades the pattern matching. The second threat is poisoning attacks which try to influence original training data to get expected result. Unlike the first threat model, this model injects malicious data into original training data to receive the adversary's desired result. As an example, the adversary sends malicious packets covertly during network traffic collection such that the learning misclassified that packet as a benign instance. Further detail explanations and examples can be found in [16] and [17].

3.2 Machine-Learning-Based Intrusion Detection Systems

A combination of two typical methods is commonly used to build an IDS such as learning or training and classification as shown in Fig. 3.2. It is difficult and expensive to obtain the bulk of labeled network connection records for supervised training in the first stage. FL might become the solution in the first place. The clustering analysis has emerged as an anomaly detection recently [6]. Clustering is an unsupervised data exploratory technique that partitions a set of unlabeled data patterns into groups or clusters such that patterns within a cluster to make similar to each other but dissimilar to other clusters' pattern [6]. FL is a tool for improving the learning process of a machine-learning algorithm. It commonly consists of feature construction, extraction, and selection. Feature construction expands the original features to enhance their expressiveness, whereas feature extraction transforms the original features into a new form and feature selection eliminates unnecessary

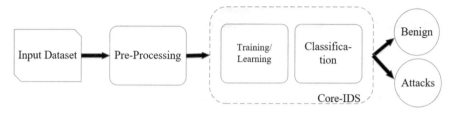

Fig. 3.2 Typical scheme IDS

features [18]. The classification task is a supervised method to distinguish benign and malicious traffics based on the provided data which usually comes from the previous step as shown in Fig. 3.2.

As can be seen in Fig. 3.2, the preprocessing step is required before entering the core-IDS module. The preprocessing module commonly consists of normalization and balancing steps. Data normalization is a process to output same value ranges of each attribute, which is vital for proper learning by any machine-learning algorithm [19]. Meanwhile, in practice, the frequency of benign traffics is much larger than the one of malicious traffics. This property could make it difficult for the core-IDS module to learn the underlying patterns correctly [20]. Therefore, the balancing process which creates the dataset with an equal ratio for both benign and malicious instances is the prerequisite step for training. However, we should use original ratio, which is unbalanced, for testing purposes to validate the IDS can be implemented in the real-world networks.

A machine-learning-based IDS has been studied for decades especially on anomaly-based IDSs. Fragkiadakis et al. [21] proposed an anomaly-based IDS using *Dempster-Shafer's rule* for measurement. Shah et al. [22] also developed an evidence theory for combining anomaly-based and misuse-based IDSs. Bostani et al. [19] proposed an anomaly-based IDS by modifying an optimum path forest combined with k-means clustering. A distributed anomaly-based IDS, called *TermID*, proposed by Kolias et al. [23] incorporates ant colony optimization and rule induction in a reduced data operation to achieve data parallelism and reduce privacy risks.

Feature selection techniques are useful to reduce the complexity of models, which leads to faster learning and real-time processing. Kayacik et al. [24] investigated the relevance of each feature in the KDD'99 Dataset with a list of the most relevant features for each class label and discussed the significance of IG rigorously. Their work confirmed the role of feature selection for building an accurate IDS model. Puthran et al. [25] also worked on relevant features in the KDD'99 Dataset and improved the Decision Tree (DT) by using binary and quad splits. Almusallam et al. [10] leveraged a filter-based feature selection method. Zaman and Karray [26] categorized IDSs based on the Transmission Control Protocol/Internet Protocol (TCP/IP) network model using a feature selection method known as the Enhanced Support Vector Decision Function (ESVDF). Louvieris et al. [27] proposed an effect-based feature identification IDS using naive Bayes as a feature selection method. Zhu et al. [28] also proposed a feature selection method using a multi-objective approach.

On the other hand, Manekar and Waghmare [29] leveraged Particle Swarm Optimization (PSO) and SVM. PSO performs feature optimization to obtain an optimized feature, after which SVM conducts the classification task. A similar approach was introduced by Saxena and Richariya [30], although the concept of weighted feature selection was introduced by Schaffernicht and Gross [31]. Exploiting SVM-based algorithms as a feature selection method was proposed by Guyon et al. [1]. This method leveraged the weights adjusted during support vector learning and resulted in ranking the importance of input features. Another related approach was proposed by Wang [32] who ranked input features based on weights learned by an ANN. This method showed the ability of DNN to find useful features among the raw data. Aljawarneh et al. [33] proposed a hybrid model of feature selection and an ensemble of classifiers which requires heavy computation.

Venkatesan et al. [34] highlighted new uses of botnets which is data theft. Botnets need two properties, stealth and resilience to steal data. This type of botnets becomes a tool for Advanced Persistent Threat (APT) agent. The authors [34] proposed a combination approach of a honeypot and a botnet intrusion detection based on RL model in a resource-constrained environment. The honeypot aims to detect intrusion occurred, while the botnet detection analyzes the behavior of bots. The RL model was developed to reduce the lifetime of stealthy botnets in a resource-constrained environment. The RL agent did not take a mission-centric approach into consideration, but the agent learned a policy to maximize the number of bots to be detected. Their RL model can be spanned during two consecutive decisions of the agent. An enterprise network containing 106 machines with 98 clients and 8 servers in 4 subnets was simulated in PeerSim software for experimental purposes. Based on the experimental results, the RL model successfully controlled the evolution of a botnet.

We have examined several feature selection methods for IDS. Huseynov et al. [35] inspected Ant Colony Clustering (ACC) method to find feature clusters of botnet traffic. The selected features in [35] are independent of traffic payload and represent the communication patterns of botnet traffic. However, this botnet detection does not scale for huge and noisy dataset due to the absence of control mechanism for clustering threshold. Kim et al. [36] tested Artificial Immune System (AIS) and swarm intelligence-based clustering to detect unknown attacks. Furthermore, Aminanto et al. [37] discussed the utility of ACA and Fuzzy Inference System (FIS) for IDS. They explored several common IDSs with a combination of learning and classification as shown in Table 3.1.

Table 3.1 Common IDSs with a combination of learning and classification

Publication	Learning	Classification
AKKK17 [37]	ACA	FIS
HKY14 [35]	ATTA-C	ATTA-C + label
KKK15 [36]	ACA	AIS
KHKY16 [38]	ACA	DT, ANN

ACA is one of the most popular clustering approaches which is originated from swarm intelligence. ACA is an unsupervised learning algorithm that can find near-optimal clustering solution without a predefined number of clusters needed [6]. However, ACA is rarely used in intrusion detection as the exclusive method for classification. Instead, ACA is combined with other supervised algorithms such as Self-Organizing Map (SOM) and support vector machine (SVM) to provide better classification result [39]. In AKKK17 [37], a novel hybrid IDS scheme based on ACA and FIS was proposed. The authors [37] applied ACA for training phase and FIS for classification phase. Then, FIS was chosen as a classification phase, because fuzzy approach can reduce the false alarm with higher reliability in determining intrusion activities [40]. Meanwhile, the same ACA with different classifiers was also examined in KKK15 [36] and KHKY16 [38] by using artificial AIS and DT as well as Artificial Neural Network (ANN), respectively. AIS is designed for the computational system and inspired by Human Inference System (HIS). AIS can differentiate between the "self" (cells that are owned by the system) and "nonself" (foreign entities to the system). ANN can learn more complex structure of certain unknown attacks due to a characteristic of ANN. Also, an improved ACA which is Adaptive Time Dependent Transporter Ants Clustering (ATTA-C) also investigated in HKY14 [35], which is one of the few algorithms that have been benchmarked on various datasets and is now publicly available under GNU agreement [35].

In addition to the abovementioned common IDSs, other IDS models were further examined by taking benefits of Hadoop framework [41] and Software-Defined Networking (SDN) environment [42]. Khalid et al. [41] proposed a method to utilize the advantages of Hadoop as well as behavioral flow analysis. This framework is particularly useful in the case of P2P traffic analysis due to inherent flow characteristics of this type of applications. Meanwhile, Lee et al. [42] proposed a novel IDS scheme that operates lightweight intrusion detection to keep a detailed analysis of attacks. In this scheme, a flow-based IDS detects intrusions but with low operating cost. When an attack is detected, the IDS requests the forwarding of attack traffic to packet-based detection so that the detailed results obtained by the packet-based detection can be analyzed later by the security experts.

References

1. I. Guyon, J. Weston, S. Barnhill, and V. Vapnik, "Gene selection for cancer classification using support vector machines," *Machine Learning*, vol. 46, no. 1–3, pp. 389–422, 2002.
2. X. Zeng, Y.-W. Chen, C. Tao, and D. van Alphen, "Feature selection using recursive feature elimination for handwritten digit recognition," in *Proc. Intelligent Information Hiding and Multimedia Signal Processing (IIH-MSP), Kyoto, Japan.* IEEE, 2009, pp. 1205–1208.
3. C. A. Ratanamahatana and D. Gunopulos, "Scaling up the naive Bayesian classifier: Using decision trees for feature selection," in *Workshop on Data Cleaning and Preprocessing (DCAP) at IEEE Int. Conf. Data Mining (ICDM), Maebashi, Japan.* IEEE, Dec 2002.
4. C. Jiang, H. Zhang, Y. Ren, Z. Han, K.-C. Chen, and L. Hanzo, "Machine learning paradigms for next-generation wireless networks," *IEEE Wireless Communications*, vol. 24, no. 2, pp. 98–105, 2017.

5. A. L. Vizine, L. N. de Castro, and E. Hrusch, "Towards improving clustering ants: an adaptive ant clustering algorithm," *Journal of Informatica*, vol. 29, no. 2, pp. 143–154, 2005.
6. C.-H. Tsang and S. Kwong, "Ant colony clustering and feature extraction for anomaly intrusion detection," *Swarm Intelligence in Data Mining*, pp. 101–123, 2006.
7. R. Rojas, "The backpropagation algorithm," *Neural Networks*. Berlin, Springer, 1996, pp. 149–182.
8. B. A. Olshausen and D. J. Field, "Sparse coding with an overcomplete basis set: A strategy employed by v1?" *Vision Research*, vol. 37, no. 23, pp. 3311–3325, 1997.
9. E. Eskin, A. Arnold, M. Prerau, L. Portnoy, and S. Stolfo, "A geometric framework for unsupervised anomaly detection," *Applications of Data Mining in Computer Security*, vol. 6, pp. 77–101, 2002.
10. N. Y. Almusallam, Z. Tari, P. Bertok, and A. Y. Zomaya, "Dimensionality reduction for intrusion detection systems in multi-data streams a review and proposal of unsupervised feature selection scheme," *Emergent Computation*, vol. 24, pp. 467–487, 2017. [Online]. Available: https://doi.org/10.1007/978-3-319-46376-6_22
11. X. Zhu and A. B. Goldberg, "Introduction to semi-supervised learning," *Synthesis lectures on artificial intelligence and machine learning*, vol. 3, no. 1, pp. 1–130, 2009.
12. Z.-H. Zhou, "A brief introduction to weakly supervised learning," *National Science Review*, 2017.
13. C. Olah, "Machine learning for humans," https://www.dropbox.com/s/e38nil1dnl7481q/machine_learning.pdf?dl=0, 2017, [Online; accessed 21-March-2018].
14. P. Laskov and R. Lippmann, "Machine learning in adversarial environments," *Machine Learning*, vol. 81, no. 2, pp. 115–119, Nov 2010. [Online]. Available: https://doi.org/10.1007/s10994-010-5207-6
15. S. J. Lewis, "Introduction to adversarial machine learning," https://mascherari.press/introduction-to-adversarial-machine-learning/, 2016, [Online; accessed 27-March-2018].
16. L. Huang, A. D. Joseph, B. Nelson, B. I. Rubinstein, and J. Tygar, "Adversarial machine learning," in *Proceedings of the 4th ACM workshop on Security and artificial intelligence*. ACM, 2011, pp. 43–58.
17. I. J. Goodfellow, J. Shlens, and C. Szegedy, "Explaining and harnessing adversarial examples," *arXiv preprint arXiv:1412.6572*, 2014.
18. H. Motoda and H. Liu, "Feature selection, extraction and construction," *Communication of IICM (Institute of Information and Computing Machinery), Taiwan*, vol. 5, pp. 67–72, 2002.
19. H. Bostani and M. Sheikhan, "Modification of supervised OPF-based intrusion detection systems using unsupervised learning and social network concept," *Pattern Recognition*, vol. 62, pp. 56–72, 2017.
20. M. Sabhnani and G. Serpen, "Application of machine learning algorithms to KDD intrusion detection dataset within misuse detection context." in *Proc. Int. Conf. Machine Learning; Models, Technologies and Applications (MLMTA), Lax Vegas, USA*, 2003, pp. 209–215.
21. A. G. Fragkiadakis, V. A. Siris, N. E. Petroulakis, and A. P. Traganitis, "Anomaly-based intrusion detection of jamming attacks, local versus collaborative detection," *Wireless Communications and Mobile Computing*, vol. 15, no. 2, pp. 276–294, 2015.
22. V. Shah and A. Aggarwal, "Enhancing performance of intrusion detection system against kdd99 dataset using evidence theory," *Int. Journal of Cyber-Security and Digital Forensics*, vol. 5(2), pp. 106–114, 2016.
23. C. Kolias, V. Kolias, and G. Kambourakis, "Termid: a distributed swarm intelligence-based approach for wireless intrusion detection," *International Journal of Information Security*, vol. 16, no. 4, pp. 401–416, 2017.
24. H. G. Kayacik, A. N. Zincir-Heywood, and M. I. Heywood, "Selecting features for intrusion detection: A feature relevance analysis on KDD 99 intrusion detection datasets," in *Proc. Privacy, Security and Trust, New Brunswick, Canada*. Citeseer, 2005.
25. S. Puthran and K. Shah, "Intrusion detection using improved decision tree algorithm with binary and quad split," in *Proc. Security in Computing and Communication*. Springer, 2016, pp. 427–438.

26. S. Zaman and F. Karray, "Lightweight IDS based on features selection and IDS classification scheme," in *Proc. Computational Science and Engineering (CSE)*. IEEE, 2009, pp. 365–370.

27. P. Louvieris, N. Clewley, and X. Liu, "Effects-based feature identification for network intrusion detection," *Neurocomputing*, vol. 121, pp. 265–273, 2013.

28. Y. Zhu, J. Liang, J. Chen, and Z. Ming, "An improved NSGA-iii algorithm for feature selection used in intrusion detection," *Knowledge-Based Systems*, vol. 116, pp. 74–85, 2017.

29. V. Manekar and K. Waghmare, "Intrusion detection system using support vector machine (SVM) and particle swarm optimization (PSO)," *Int. Journal of Advanced Computer Research*, vol. 4, no. 3, pp. 808–812, 2014.

30. H. Saxena and V. Richariya, "Intrusion detection in KDD99 dataset using SVM-PSO and feature reduction with information gain," *Int. Journal of Computer Applications*, vol. 98, no. 6, 2014.

31. E. Schaffernicht and H.-M. Gross, "Weighted mutual information for feature selection," in *Proc. Artificial Neural Networks, Espoo, Finland*. Springer, 2011, pp. 181–188.

32. Z. Wang, "The applications of deep learning on traffic identification," in *Conf. BlackHat, Las Vegas, USA*. UBM, 2015.

33. S. Aljawarneh, M. Aldwairi, and M. B. Yassein, "Anomaly-based intrusion detection system through feature selection analysis and building hybrid efficient model," *Journal of Computational Science*, Mar 2017. [Online]. Available: http://dx.doi.org/10.1016/j.jocs.2017.03.006

34. S. Venkatesan, M. Albanese, A. Shah, R. Ganesan, and S. Jajodia, "Detecting stealthy botnets in a resource-constrained environment using reinforcement learning," in *Proceedings of the 2017 Workshop on Moving Target Defense*. ACM, 2017, pp. 75–85.

35. K. Huseynov, K. Kim, and P. Yoo, "Semi-supervised botnet detection using ant colony clustering," in *Symp. Cryptography and Information Security (SCIS), Kagoshima, Japan*, 2014.

36. K. M. Kim, H. Kim, and K. Kim, "Design of an intrusion detection system for unknown-attacks based on bio-inspired algorithms," in *Computer Security Symposium (CSS), Nagasaki, Japan*, 2015.

37. M. E. Aminanto, H. Kim, K. M. Kim, and K. Kim, "Another fuzzy anomaly detection system based on ant clustering algorithm," *IEICE Transactions on Fundamentals of Electronics, Communications and Computer Sciences*, vol. 100, no. 1, pp. 176–183, 2017.

38. K. M. Kim, J. Hong, K. Kim, and P. Yoo, "Evaluation of ACA-based intrusion detection systems for unknown-attacks," in *Symp. on Cryptography and Information Security (SCIS)*, Kumamoto, Japan, 2016.

39. C. Kolias, G. Kambourakis, and M. Maragoudakis, "Swarm intelligence in intrusion detection: A survey," *Computers & Security*, vol. 30, no. 8, pp. 625–642, 2011.

40. A. Karami and M. Guerrero-Zapata, "A fuzzy anomaly detection system based on hybrid PSO-Kmeans algorithm in content-centric networks," *Neurocomputing*, vol. 149, pp. 1253–1269, 2015.

41. K. Huseynov, P. D. Yoo, and K. Kim, "Scalable P2P botnet detection with threshold setting in Hadoop framework," *Journal of the Korea Institute of Information Security and Cryptology*, vol. 25, no. 4, pp. 807–816, 2015.

42. D. S. Lee, "Improving detection capability of flow-based IDS in SDN," *KAIST, MS. Thesis*, 2015.

Chapter 4
Deep Learning

Abstract This chapter defines a brief history and definition of deep learning. Due to a variety of models belonging to deep learning, we classify deep learning models into a tree which has three branches: generative, discriminative, and hybrid. In each model, we show some learning model examples in order to see the difference among three models.

4.1 Classification

Deep learning originally comes from the advancements of NN algorithm. Various methods have been applied in order to overcome the limitations of one hidden layer only in NN. Those methods employ consecutive hidden layers which are hierarchically cascaded. Due to a variety of models belonging to deep learning, Aminanto et al. [1] classified several deep learning models based on approaches as guided by Deng [2, 3] which differentiates deep learning into three subgroups, generative, discriminative, and hybrid. The classification is based on the intention of architectures and techniques, e.g., synthesis/generation, or recognition/classification. The classification of the deep learning methods is shown in Fig. 4.1.

4.2 Generative (Unsupervised Learning)

Unsupervised learning or so-called generative model uses unlabeled data. The main concept of applying the generative architectures to the pattern recognition is unsupervised learning or pre-training [2]. Since learning the lower levels of subsequent networks are difficult, deep generative structures are needed. Thus, from the limited training data, learning each lower layer in layer-by-layer approach without relying on all the upper layers is essential. The generative models also intend to learn joint statistical distributions of given data [3]. These models calculate

© The Author(s), under exclusive license to Springer Nature Singapore Pte Ltd. 2018   27
K. Kim et al., *Network Intrusion Detection using Deep Learning*,
SpringerBriefs on Cyber Security Systems and Networks,
https://doi.org/10.1007/978-981-13-1444-5_4

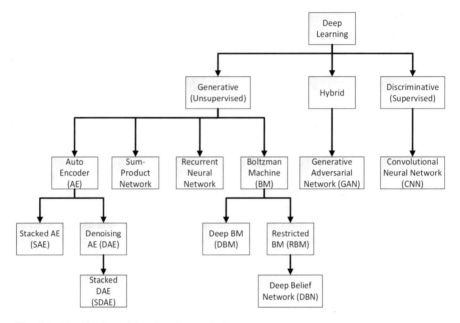

Fig. 4.1 Classification of deep learning methods

the joint probability given the input and choose the class label with the highest probability [4]. There are a number of methods that are classified as unsupervised learning.

4.2.1 Stacked (Sparse) Auto-Encoder

(Sparse) AE can be used as deep learning technique by an unsupervised greedy layer-wise pre-training algorithm known as SAE. Here, pre-training refers to the training of a single AE using a single hidden layer. Each AE is trained separately before being cascaded afterward. This pre-training phase is required to construct a stacked AE. In this algorithm, all layers except the output layer are initialized in a multilayer NN. Each layer is then trained in an unsupervised manner as an AE, which constructs new representations of the input.

SAE is trained with the same neuron number of both input and output layers. Meanwhile, the nodes in the hidden layer are lesser than the input which represents a new less feature set. This architecture leads to new capability that can reconstruct the data after complicated computations. AE aims to learn a compact set of data efficiently and can be stacked to build a deep network. Training results of each hidden layer are cascaded, which can provide new transformed features by different depths. To train more precisely, one can append an additional classifier layer with class labels [5]. Besides, a Denoising Auto-Encoder (DAE) is trained to reconstruct

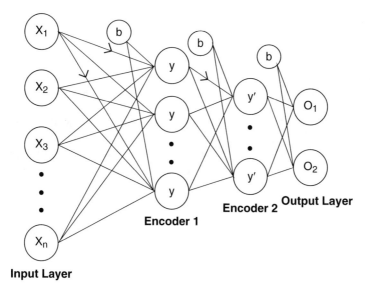

Fig. 4.2 SAE network with two hidden layers and two target classes

a precise correction input from corrupted by noise input [6]. The DAE may also be stacked to build deep networks as well.

The performance of the unsupervised greedy layer-wise pre-training algorithm can be significantly more accurate than the supervised one. This is because the greedy supervised procedure may behave too greedy as it extracts less information and considers one layer only [7, 8]. A NN containing only one hidden layer may discard some of the information about the input data since more information could be exploited by composing additional hidden layers. The features from the pre-training phase, which is greedy layer-wise, can be used either as an input to a standard supervised machine-learning algorithm or as initialization for a deep supervised NN.

Figure 4.2 shows the SAE network with two hidden layers and two target classes. The final layer implements the softmax function for the classification in the DNN. This softmax function makes SAE both unsupervised and supervised learning. Softmax function is a generalized term of the logistic function that suppresses the K-dimensional vector $\mathbf{v} \in \mathbb{R}^K$ into K-dimensional vector $\mathbf{v}^* \in (0, 1)^K$, which adds up to 1. In this function, T and C are defined as the number of training instances and the number of classes, respectively. The softmax layer minimizes the loss function, which is either the cross-entropy function like Eq. (4.1) or the mean-squared error.

$$E = \frac{1}{T} \sum_{j=1}^{T} \sum_{i=1}^{C} \left[z_{ij} \log y_{ij} + (1 - z_{ij}) \log \left(1 - y_{ij} \right) \right], \qquad (4.1)$$

4.2.2 Boltzmann Machine

BM is a network of binary units that symmetrically paired [9], which means all input nodes are linked to all hidden nodes. BM is a shallow model with one hidden layer only. BM has a structure of neuron units that make stochastic decisions about whether active or not [3]. If one BM output is cascaded into multiple BMs, it is called deep BM (DBM). Meanwhile, RBM is a customized BM without connections between the hidden nodes and input nodes too [9]. RBM consists of visible and hidden variables such that their relations can be figured out. Visible here means neurons in input which is training data. If multiple layers of RBM are stacked, a layer-by-layer scheme is called deep belief network (DBN). DBN could be used as a feature extraction method for dimensionality reduction when unlabeled dataset and back-propagation are used (which means unsupervised training). In contrast, DBN is used for classification when appropriately labeled dataset with feature vectors are used (which means supervised training) [10].

4.2.3 Sum-Product Networks

Another deep generative model is Sum-Product Networks (SPN), which is a directed acyclic graph with variables as leaves, sum and product operations as internal nodes, and weighted edges [11]. The sum nodes provide mixture models, while the product nodes express the feature hierarchy [3]. Therefore, we can consider SPN as a combination of mixture models and feature hierarchies.

4.2.4 Recurrent Neural Networks

Recurrent Neural Networks (RNN) is an extension of neural networks with cyclic links to process sequential information. This cyclic links placed between higher- and lower-layer neurons which enable RNN to propagate data from previous to current events. This property makes RNN having a memory of time series events [12]. Figure 4.3 shows a single loop of RNN on the left side which is comparable to the right-side topology when the loop is broken.

One advantage of RNN is the ability to connect previous information to present task; however, it cannot reach "far" previous memory. This problem is commonly known as long-term dependencies. Long Short-Term Memory (LSTM) networks are introduced by Hochreiter and Schmidhuber [14] to overcome this problem. LSTMs are an extension of RNN with four neural networks in a single layer, where RNN have one only as shown in Fig. 4.4.

The main advantage of LSTM is the existence of state cell which the line is passing through in the top of every layer. The cell accounts for propagating information from the previous layer to the next one. Then, "gates" in LSTM would

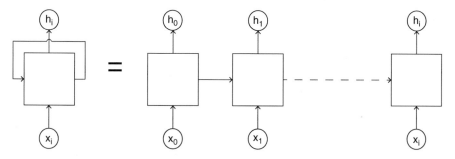

Fig. 4.3 RNN with unrolled topology [13]

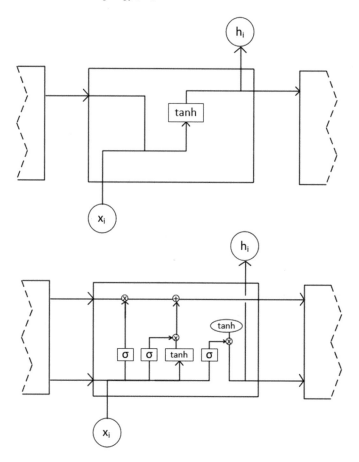

Fig. 4.4 RNN topology (top) vs LSTM topology (bottom) [13]

manage which information will be passed or dropped. There are three gates to control the information flow, namely, input, forget, and output gates [15]. These gates are composed of a sigmoid NN and an operator as shown in Fig. 4.4.

4.3   Discriminative

Supervised learning or discriminative model is intended to distinguish some parts of data for pattern classification with labeled data [2]. An example of the discriminative architecture is CNN which employs a special architecture particularly suitable for image recognition. The main advantage of CNN is that the handcrafted feature extraction is not necessary. CNN can train multilayer networks with gradient descent to learn complex, high-dimensional, nonlinear mappings from large collections of data [16]. CNN uses three basic concepts: local receptive fields, shared weights, and pooling [17]. One extensive research that successfully deployed using CNN is AlphaGo by Google [18]. Other examples of discriminative models are linear and logistic regressions [4].

RNN can also be considered as a discriminative model when the output of RNN is used as label sequences for the input [3]. One example of this network was proposed by Graves [19] who leveraged RNNs to build a probabilistic sequence transduction system, which can transform any input sequence into any finite, discrete output sequence.

4.4   Hybrid

The hybrid deep architecture combines both generative and discriminative architectures. The hybrid structure aims to distinguish data as well as discriminative approach. However, in the early step, it has assisted in a significant way with the generative architectures results. An example of hybrid architecture is Deep Neural Network (DNN). However, some confusion terms between DNN and DBN happens. In the open literature, DBN also uses backpropagation discriminative training as a "fine-tuning." This concept of DBN is similar DNN [2]. According to Deng [3], DNN which is defined as a multilayer network with cascaded fully connected hidden layers, uses stacked RBM as a pre-training phase. Many other generative models can be considered as discriminative or hybrid models when the classification task is added with the class labels.

4.4.1   Generative Adversarial Networks (GAN)

Goodfellow [20] introduced a novel framework which trains both generative and discriminative models at the same time, which the generative model G captures the data distribution and the discriminative model D distinguishes the original input data and the data coming from the model G. It is a zero-sum game of G and D models [4] where model G aims to counterfeit the original input data, while model D aims to discriminate the original input and output of model G. According to Dimokranitou

Fig. 4.5 Typical architecture
of GAN

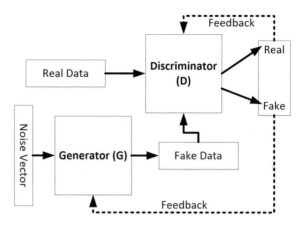

[4], the advantage of GAN are to keep consistency after the equilibrium was achieved, no approximate inference or Markov chains are needed, and can be trained with missing or limited data. On the other hand, the disadvantage of applying GAN is to find the equilibrium between G and D models. A typical architecture of GAN is shown in Fig. 4.5.

References

1. M. E. Aminanto and K. Kim, "Deep learning in intrusion detection system: An overview," *International Research Conference on Engineering and Technology 2016*, Bali, Indonesia, June 28–30, 2016.
2. L. Deng, "A tutorial survey of architectures, algorithms, and applications for deep learning," *APSIPA Transactions on Signal and Information Processing*, vol. 3, 2014.
3. L. Deng, D. Yu, *et al.*, "Deep learning: methods and applications," *Foundations and Trends® in Signal Processing*, vol. 7, no. 3–4, pp. 197–387, 2014.
4. A. Dimokranitou, "Adversarial autoencoders for anomalous event detection in images," Ph.D. dissertation, Purdue University, 2017.
5. Z. Wang, "The applications of deep learning on traffic identification," in *Conf. BlackHat, Las Vegas, USA*. UBM, 2015.
6. P. Vincent, H. Larochelle, I. Lajoie, Y. Bengio, and P.-A. Manzagol, "Stacked denoising autoencoders: Learning useful representations in a deep network with a local denoising criterion," *Journal of Machine Learning Research*, vol. 11, no. Dec, pp. 3371–3408, 2010.
7. Y. Bengio, P. Lamblin, D. Popovici, H. Larochelle, *et al.*, "Greedy layer-wise training of deep networks," *Advances in neural information processing systems*, vol. 19, pp. 153–160, Sep 2007.
8. Y. Bengio, Y. LeCun, *et al.*, "Scaling learning algorithms towards AI," *Large-scale kernel machines*, vol. 34, no. 5, pp. 1–41, 2007.
9. R. Salakhutdinov and G. Hinton, "Deep boltzmann machines," *Artificial Intelligence and Statistics*, pp. 448–455, 2009.
10. M. Salama, H. Eid, R. Ramadan, A. Darwish, and A. Hassanien, "Hybrid intelligent intrusion detection scheme," *Soft computing in industrial applications*, pp. 293–303, 2011.
11. H. Poon and P. Domingos, "Sum-product networks: A new deep architecture," in *Computer Vision Workshops (ICCV Workshops), 2011 IEEE International Conference on*. IEEE, 2011, pp. 689–690.

12. R. C. Staudemeyer, "Applying long short-term memory recurrent neural networks to intrusion detection," *South African Computer Journal*, vol. 56, no. 1, pp. 136–154, 2015.
13. C. Olah, "Understanding LSTM networks," http://colah.github.io/posts/2015-08-Understanding-LSTMs/, 2015, [Online; accessed 20-February-2018].
14. S. Hochreiter and J. Schmidhuber, "Long short-term memory," *Neural computation*, vol. 9, no. 8, pp. 1735–1780, 1997.
15. J. Kim, J. Kim, H. L. T. Thu, and H. Kim, "Long short term memory recurrent neural network classifier for intrusion detection," in *Platform Technology and Service (PlatCon), 2016 International Conference on*. IEEE, 2016, pp. 1–5.
16. Y. LeCun, L. Bottou, Y. Bengio, and P. Haffner, "Gradient-based learning applied to document recognition," *Proceedings of the IEEE*, vol. 86, no. 11, pp. 2278–2324, 1998.
17. M. A. Nielsen, "Neural networks and deep learning," 2015.
18. D. Silver, A. Huang, C. J. Maddison, A. Guez, L. Sifre, G. Van Den Driessche, J. Schrittwieser, I. Antonoglou, V. Panneershelvam, M. Lanctot, *et al.*, "Mastering the game of Go with deep neural networks and tree search," *Nature*, vol. 529, no. 7587, pp. 484–489, 2016.
19. A. Graves, "Sequence transduction with recurrent neural networks," *arXiv preprint arXiv:1211.3711*, 2012.
20. I. Goodfellow, J. Pouget-Abadie, M. Mirza, B. Xu, D. Warde-Farley, S. Ozair, A. Courville, and Y. Bengio, "Generative adversarial nets," in *Advances in neural information processing systems*, 2014, pp. 2672–2680.

Chapter 5
Deep Learning-Based IDSs

Abstract This chapter reviews recent IDSs leveraging deep learning models as their methodology which were published during 2016 and 2017. The critical issues like problem domain, methodology, dataset, and experimental result of each publication will be discussed. These publications can be classified into three different categories according to deep learning classification in Chap. 4, namely, generative, discriminative, and hybrid. The generative model group consists of IDSs that use deep learning models for feature extraction only and use shallow methods for the classification task. The discriminative model group contains IDSs that use a single deep learning method for both feature extraction and classification task. The hybrid model group includes IDSs that use more than one deep learning method for generative and discriminative purposes. All IDSs are compared to overview the advancement of deep learning in IDS researches.

5.1 Generative

This sub-chapter groups IDSs that use deep learning for feature extraction only and use shallow methods for the classification task.

5.1.1 Deep Neural Network

Roy et al. [1] proposed an IDS by leveraging deep learning models and validated that a deep learning approach can improve IDS performance. DNN is selected comprising of multilayer feedforward NN with 400 hidden layers. Shallow models, rectifier and softmax activation functions, are used in the output layer. Two advantages of feedforward neural network are to provide a precise approximation for complex multivariate nonlinear function directly from input values and to give the robust modeling for large classes. Besides that, the authors claimed that DNN

© The Author(s), under exclusive license to Springer Nature Singapore Pte Ltd. 2018   35
K. Kim et al., *Network Intrusion Detection using Deep Learning*,
SpringerBriefs on Cyber Security Systems and Networks,
https://doi.org/10.1007/978-981-13-1444-5_5

is better than DBN since the discriminating power by characterizing the posterior distributions of classes is suitable for the pattern classification [1].

For validation, KDD Cup'99 dataset was used. This dataset has 41 features that was given as the input to the network. The authors divided all the training data into 75% for training and 25% for validation. They also compared the performance of a shallow classifier, SVM. Based on their experimental result, DNN outperforms SVM by the accuracy of 99.994%, while SVM achieved 84.635% only. This result showed the effectiveness of DNN for IDS purposes.

5.1.2   Accelerated Deep Neural Network

Another DNN but different architecture was proposed by Potluri and Diedrich [2] in 2016. This paper mainly focuses on improving DNN implementation for IDS by using multi-core CPUs and GPUs. This is important since DNN requires large computation for training [3]. They reviewed some IDSs utilizing a hardware enhancement: GPU, multicore CPU, memory management, and FPGA. Also, a way of load balancing (and splitting) or parallel processing was discussed. A deep learning model, SAE was chosen to construct the DNN in this work. The architecture of this network has 41 input features from NSL-KDD dataset, 20 neurons in the first hidden layer by first AE, 10 neurons in the second hidden layer of the second AE, and 5 neurons in the output layer containing softmax activation function. In the training phase, each AE is trained separately but in sequence since the hidden layer of the first AE becomes the input of second AE. There are two times of fine-tuning processes, the first one done by softmax activation function and the second one done by backpropagation through the entire network.

NSL-KDD dataset was selected for testing this approach. This dataset is a revised version of KDD Cup'99 dataset. It has the same number of features which is 41 but with more rational distributions and without redundant instances existing in KDD Cup'99 dataset. The authors firstly tested the network with different attack class combinations from two classes to four classes. The lesser number of attack classes performs better than the higher number of attack classes as expected since the imbalance class distribution leads to a good result for fewer attack types. For the acceleration, the authors used two different CPUs and a GPU. They also experimented using serial and parallel CPUs. Their experimental result shows that the training using parallel CPU achieved three times faster than using serial CPU. The training using GPU delivered similar performance to parallel CPU as well. An interesting point here is the training using parallel of the second CPU is faster than GPU. They explained that this case happens due to the clock speed of the current CPU which is too high. Unfortunately, the authors do not provide performance comparison regarding detection accuracy or false alarm rate.

5.1.3 Self-Taught Learning

Self-Taught Learning (STL) was proposed as a deep learning model for IDS by Niyaz et al. [4]. The authors mentioned two challenges to develop an efficient IDS. The first challenge is to select feature since the selected features for a particular attack might be different for other attack types. The second challenge is to deal with the limited amounts of a labeled dataset for training purpose. Therefore, a generative deep learning model was chosen in order to deal with this unlabeled dataset. The proposed STL comprises of two stages, Unsupervised Feature Learning (UFL) and Supervised Feature Learning (SFL). For UFL, the authors leveraged sparse AE while softmax regression for SFL. Figure 5.1 shows the two-stage process of STL used in this paper. The UFL accounts for feature extraction with unlabeled dataset, while the SFL accounts for classification task with labeled data.

The authors verified their approach using NSL-KDD dataset. Before the training process, the authors defined a preprocessing step for the dataset which contains 1-to-N encoding and min-max normalization. After 1-to-N encoding process, 121 features were ready for normalization step and input features for the UFL. Tenfold cross-validation and test dataset from NSL-KDD dataset were selected for training and test data, respectively. The authors also evaluated the STL for three different attack combinations, 2-class, 5-class, and 23-class. In general, their STL achieved higher than 98% of classification accuracy for all combinations during the training phase. In the testing phase, the STL achieved an accuracy of 88.39% and 79.10% for 2-class and 5-class classifications, respectively. They mentioned that the future work is to develop a real-time IDS using deep learning models on raw network traffic.

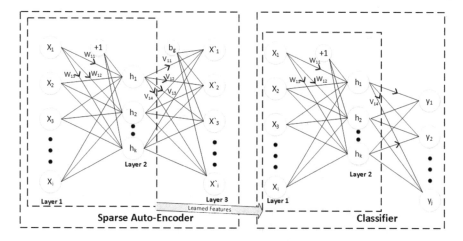

Fig. 5.1 The two stages of STL [4]

5.1.4   Stacked Denoising Auto-Encoder

Yu et al. [5] introduced a session-based IDS using a deep learning architecture. They came up with common IDS shortcomings: high FP and false negative. Most attack features in common dataset are heavily structured and have special semantics involved in specific expert knowledge, and the heavily handcrafted dataset is closely related to particular attack classes. Therefore, a deep learning model was leveraged since unsupervised deep learning can learn essential features automatically from large data. The proposed approach consists of extracting features from raw data and applying unsupervised SDAE. Session-based data was extracted from the raw network data, which was drawn from UNB ISCX 2012 and CTU-13 for benign and botnet instances, respectively. Since the data is extracted from the raw data, a preprocessing step is necessary. Data preprocessing procedure consists of session construction, record construction, and normalization. Session construction distinguishes three different sessions, namely, TCP, UDP, and ICMP. Record construction draws first 17 features from packet headers and the rest 983 features from the payload. In the end, normalization session uses the min-max function. The SDAE itself contains two hidden layers and a softmax regression layer. For the denoising purpose, the authors randomly set 10%, 20%, 30% of the input features by zero. The authors mentioned that the advantages of using SDAE are learning capability of important features from unlabeled instances automatically, denoising strategy of making robustness from missing and noisy input, and good dimensionality reduction when the hidden layer is nonlinear.

They measured accuracy, precision, recall, F-score, and ROC curve as the performance metrics. Binary and multi-class classifications were used along with 43% of the dataset and whole dataset combinations to verify the performance of SDAE. The SDAE also was compared to other deep learning models, namely, SAE, DBN, and AE-CNN models. In overall, the SDA achieved the best performance with the highest accuracy rate of 98.11% of multi-class classification using the whole dataset.

5.1.5   Long Short-Term Memory Recurrent Neural Network

Kim et al. [6] adopted the generative approach of LSTM-RNN for an IDS purpose. They leveraged softmax regression layer as the output layer. Other hyper-parameters are 50, 100, and 500 of batch size, time step, and epoch, respectively. Also, Stochastic Gradient Descent (SGD) and MSE were used as the optimizer and loss function, respectively. 41 input features were drawn from KDD Cup'99 dataset. Experimental results show the best learning rate is 0.01 and hidden layer size is 80 with 98.88% of DR and 10.04% of false alarm rate. Similar network topology was also proposed by Liu et al. [7] with different hyper-parameters: time step, batch size, and age are 50, 100, and 500, respectively. Using KDD Cup'99 dataset, 98.3% of DR and 5.58% of false alarm rate were achieved.

5.2 Discriminative

This sub-chapter groups IDSs that use a single deep learning method for both feature extraction and classification task.

5.2.1 Deep Neural Network in Software-Defined Networks

SDN is an emerging network technology of current applications since it has a unique property built by the controller plane and data plane. The controller plane decouples the network control and forwarding functions. The centralized approach of controller plane makes SDN controller suitable for IDS function due to the whole network captured by the controller. Unfortunately, due to the separation of control and data planes, it leads to some critical threats. Tang et al. [8] proposed a DNN approach for IDS in SDN context. The DNN architecture is 6-12-6-3-2 which means 6 input features; hidden layers with 12, 6, and 3 neurons for each; and 2 classes of output as shown in Fig. 5.2.

They used NSL-KDD dataset to verify their approach. Since the dataset has 41 features, the authors selected 6 fundamental features in SDN based on their expertise. They measured accuracy, precision, recall, F-score, and ROC curve as the performance metrics. From their experiments, the learning rate of 0.001 is found to be the best hyper-parameter since the learning rate of 0.0001 is meant to be over-fitted. The proposed approach was then compared to the previous work which leverages a variety of machine-learning models. The DNN achieved 75.75% of accuracy, which is lower than other methods using whole 41 features but higher than other methods using 6 features only. From this fact, the authors claimed that

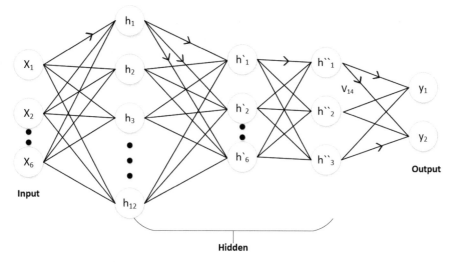

Fig. 5.2 Architecture of DNN used in [8]

the proposed DNN could generalize and abstract the characteristics of network traffic with limited of features alone.

5.2.2   Recurrent Neural Network

Yin et al. [9] highlighted the shortcoming of traditional machine-learning method-ologies which is unable to solve the classification from the massive intrusion data classification problem efficiently. They took advantages of RNN implementation in IDS context. RNN contains forward and backward propagations, where the latter is the same neural network which computes the residual of forwarding propagation. The proposed RNN-IDS begins with data preprocessing step which comprises of numericalization and normalization. Feature-ready data are propagated to training step of RNN. The output model from the training is used to apply for the testing step using the test dataset.

For experimental purposes, they used NSL-KDD dataset used both for training and testing. The original features are 41 features but became 122 features after numericalization which maps string to binary. Two types of classification were tested, namely, binary and multi-class classification. Based on the experimental results, the best hyper-parameters for the binary classification are learning rate of 0.1, an epoch of 100, and hidden nodes of 80 with an accuracy of 83.28% (using KDDTest$^+$). Meanwhile, for the multi-class classification the best hyper-parameters are learning rate of 0.5 and hidden nodes of 80 with an accuracy of 81.29%. The RNN-IDS outperformed other machine-learning methodologies tested by the authors for both binary and multi-class classification.

5.2.3   Convolutional Neural Network

Li et al. [10] experimented using CNN for the feature extractor and classifier in IDS's. CNN achieved many successful implementations in image-related classifica-tion tasks; however, it is still a big challenge for text classification. Therefore, the main challenge of implementing CNN in IDS context is the image conversion step, which is proposed by Li et al. [10]. NSL-KDD dataset was used for experimental purposes. The image conversion step begins by mapping of 41 original features into 464 binary vectors. The mapping step comprises of two types of mapping, one hot encoder for symbolic features and the other hot encoder for continuous features with ten binary vectors. The image conversion step continues with converting 464 vectors into 8 × 8 pixel images. These images are ready for training input of CNN. The authors decided to experiment with learned CNN models, ResNet 50 and GoogLeNet. Experimental results on KDDTest$^+$ show the accuracy of 79.14% and 77.14% using ResNet 50 and GoogLeNet, respectively. Although this result does not improve the state of the art of IDS, this work demonstrated how to apply CNN with image conversion in IDS context.

5.2.4 *Long Short-Term Memory Recurrent Neural Network*

LSTM-RNN became more popular due to its successful applications in various research areas. The capability of self-learning from the previous events can be applied for IDS which means the learning step from the previous attack behaviors. Some IDSs that were implementing LSTM-RNN are described as follows.

5.2.4.1 LSTM-RNN Staudemeyer

Staudemeyer [11] experimented various network topologies of LSTM-RNN for network traffic modeling as a time series. Training data was extracted from KDD Cup'99 dataset. The author also selected subset of salient features by using decision tree algorithm and compared to the whole and subset features performance in the experiment. Their experiments were run with different parameters and structures of an LSTM-RNN, such as the number of memory blocks and the cells per memory block, the learning rate, and the number of passes through the data. Besides that, experiments were also executed with a layer of hidden neurons, with peephole connections, with forget gates, and with LSTM shortcut connections. Based on the experimental results, the best performance was achieved by four memory blocks containing two cells, with forget gates and shortcut connections, 0.1 of learning rate and up to 1,000 epochs. The overall accuracy is 93.82%. They also mentioned in the conclusion that LSTM-RNN is suitable for classifying attacks with a big number of records and poor for a limited number of attack instances.

5.2.4.2 LSTM-RNN for Collective Anomaly Detection

Bontemps et al. [12] leveraged LSTM-RNN in IDS for two objectives: a time series anomaly detector and collective anomaly detector by proposing a circular array. Collective anomaly itself is a collection of related anomalous data instances concerning the whole dataset [12]. They used KDD Cup'99 dataset for their experiment and explained the preprocessing steps needed to build a time series dataset from KDD Cup'99 dataset.

5.2.4.3 GRU in IoT

Putchala [13] implemented a simplified form LSTM, called Gated Recurrent Unit (GRU) in IoT environments. GRU is suitable for IoT due to its simplicity which makes to reduce a number of gates in the network. GRU merges both forget and input gates to an update gate and combines the hidden and cell states to be a simple structure as shown in Fig. 5.3.

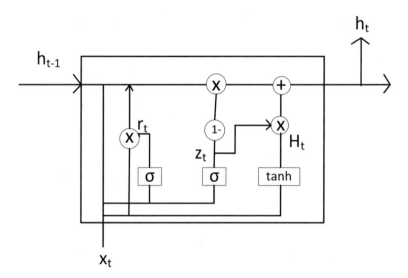

Fig. 5.3 GRU cell [13]

The author then adopted a multilayer GRU, which is GRU cells used in each hidden layer of RNN and feature selection also done by using random forest algorithm. The experiments were conducted using KDD Cup'99 dataset and achieved 98.91% and 0.76% of accuracy and false alarm rate, respectively.

5.2.4.4   LSTM-RNN for DDoS

Bediako [14] proposed a Distributed Denial of Service (DDoS) detector using LSTM-RNN. The author checked the performance of LSTM-RNN using both CPU and GPU. NSL-KDD dataset was used for experiments. The notable detection accuracy is 99.968%.

5.3   Hybrid

This sub-chapter includes an IDS that uses more than one deep learning model for generative and discriminative purposes.

5.3.1   Adversarial Networks

Dimokranitou et al. [15] proposed an abnormal events detector in images using an adversarial networks. Although the detector was not an IDS, it has the same objective to detect anomalies. They implemented an adversarial AE which combines

AEs and GAN. The network attempts to match the aggregated posterior of the hidden code vector of AE, with an arbitrary prior distribution. The reconstruction error of learned AE is low for normal events and high for irregular events.

5.4 Deep Reinforcement Learning

Choi and Cho [16] highlighted two main problems of intrusion detection in database applications where the number of benign data is much larger than malicious data in the real environment, and the internal intrusion is more difficult to detect. The proposed approach was an adaptive IDS for database applications using an Evolutionary Reinforcement Learning (ERL) which combines the evolutionary learning and the reinforcement learning for the learning process of a population and an individual, respectively. The approach comprises of two MLPs, a behavior and an evaluation network. The former network aims to detect abnormal query by performing the error backpropagation for the reinforcement learning depending on the weights of the network, while the latter network provides a learning rate as feedback to improve the detection rate of the behavior network. Since the evaluation network used for evolutionary learning, the network evolves to explore optimal model. Experiments were done using a particular scenario, called TPC-E, which is an online transaction processing workload of a brokerage company [16]. A 90% of classification accuracy was achieved after 25 generations.

Feng and Xu [17] concerned about detecting unknown attacks in Cyber-Physical System (CPS). Then a novel deep RL-based optimal strategy was proposed [17]. The novelty came as an explicit cyber state-dependent dynamics and a model of the zero-sum game to solve the Hamiltonian-Jacobi-Isaac (HJI) equation. A deep RL algorithm with game theoretical actor critic NN structure was developed to address the HJI equation. The deep RL network consists of three multilayer NNs; the first is used in critic part, the second is used to approximate the possible worst attack policy, and the last is used to estimate the optimal defense policy in real time.

5.5 Comparison

We compare and summarize all previous work mentioned earlier in this chapter. All the discussed models using KDD Cup'99 and NSL-KDD datasets were summarized in Tables 5.1 and 5.2, respectively.

The overall performances of IDSs on KDD Cup'99 are promising, as expected, more than 90% of accuracy. Four IDSs in Table 5.1 are using LSTM-RNN approach which means that a time series analysis is suitable for distinguishing benign and anomalies in network traffic. Even more, GRU [13] demonstrated that a lightweight deep learning model is very useful to be implemented in IoT environments which is of crucial importance these days.

Table 5.1 Model comparisons on KDD Cup'99 dataset

Model	Feature Extractor	Classifier	Accuracy (%)
DNN [1]	FF-NN	Softmax	99.994
LSTM-RNN-K [6]	LSTM-RNN	Softmax	96.930
LSTM-RNN-L [7]	LSTM-RNN	Softmax	98.110
LSTM-RNN-S [11]	LSTM-RNN	LSTM-RNN	93.820
GRU [13]	GRU	GRU	98.920

Table 5.2 Model comparisons on NSL KDD dataset

Model	Feature Extractor	Classifier	Accuracy (%)
STL [4]	AE	Softmax	79.10
DNN-SDN [8]	NN	NN	75.75
RNN [9]	RNN	RNN	81.29
CNN [10]	CNN	CNN	79.14

There is still a space for improvement when we are using NSL-KDD dataset as shown in Table 5.2. The most accurate model is RNN [9] with 81.29% of accuracy. Again, this fact infers that a time series analysis may improve IDS performance. Although IDS using CNN has not achieved the best performance, by applying a proper text-to-image conversion, we may gain the benefit of CNN which was previously shown to be the best for the image recognition.

References

1. S. S. Roy, A. Mallik, R. Gulati, M. S. Obaidat, and P. Krishna, "A deep learning based artificial neural network approach for intrusion detection," in *International Conference on Mathematics and Computing*. Springer, 2017, pp. 44–53.
2. S. Potluri and C. Diedrich, "Accelerated deep neural networks for enhanced intrusion detection system," in *Emerging Technologies and Factory Automation (ETFA), 2016 IEEE 21st International Conference on*. IEEE, 2016, pp. 1–8.
3. H. Larochelle, Y. Bengio, J. Louradour, and P. Lamblin, "Exploring strategies for training deep neural networks," *Journal of machine learning research*, vol. 10, no. Jan, pp. 1–40, 2009.
4. A. Javaid, Q. Niyaz, W. Sun, and M. Alam, "A deep learning approach for network intrusion detection system," in *Proceedings of the 9th EAI International Conference on Bio-inspired Information and Communications Technologies (formerly BIONETICS)*. ICST (Institute for Computer Sciences, Social-Informatics and Telecommunications Engineering), 2016, pp. 21–26.
5. Y. Yu, J. Long, and Z. Cai, "Session-based network intrusion detection using a deep learning architecture," in *Modeling Decisions for Artificial Intelligence*. Springer, 2017, pp. 144–155.
6. J. Kim, J. Kim, H. L. T. Thu, and H. Kim, "Long short term memory recurrent neural network classifier for intrusion detection," in *Platform Technology and Service (PlatCon), 2016 International Conference on*. IEEE, 2016, pp. 1–5.
7. Y. LIU, S. LIU, and Y. WANG, "Route intrusion detection based on long short term memory recurrent neural network," *DEStech Transactions on Computer Science and Engineering*, no. cii, 2017.

8. T. A. Tang, L. Mhamdi, D. McLernon, S. A. R. Zaidi, and M. Ghogho, "Deep learning approach for network intrusion detection in software defined networking," in *Wireless Networks and Mobile Communications (WINCOM), 2016 International Conference on*. IEEE, 2016, pp. 258–263.
9. C. Yin, Y. Zhu, J. Fei, and X. He, "A deep learning approach for intrusion detection using recurrent neural networks," *IEEE Access*, vol. 5, pp. 21 954–21 961, 2017.
10. Z. Li, Z. Qin, K. Huang, X. Yang, and S. Ye, "Intrusion detection using convolutional neural networks for representation learning," in *International Conference on Neural Information Processing*. Springer, 2017, pp. 858–866.
11. R. C. Staudemeyer, "Applying long short-term memory recurrent neural networks to intrusion detection," *South African Computer Journal*, vol. 56, no. 1, pp. 136–154, 2015.
12. L. Bontemps, J. McDermott, N.-A. Le-Khac, *et al.*, "Collective anomaly detection based on long short-term memory recurrent neural networks," in *International Conference on Future Data and Security Engineering*. Springer, 2016, pp. 141–152.
13. M. K. Putchala, "Deep learning approach for intrusion detection system (ids) in the internet of things (iot) network using gated recurrent neural networks (gru)," Ph.D. dissertation, Wright State University, 2017.
14. P. K. Bediako, "Long short-term memory recurrent neural network for detecting DDoS flooding attacks within tensorflow implementation framework." 2017.
15. A. Dimokranitou, "Adversarial autoencoders for anomalous event detection in images," Ph.D. dissertation, Purdue University, 2017.
16. S.-G. Choi and S.-B. Cho, "Adaptive database intrusion detection using evolutionary reinforcement learning," in *International Joint Conference SOCO'17-CISIS'17-ICEUTE'17 León, Spain, September 6–8, 2017, Proceeding*. Springer, 2017, pp. 547–556.
17. M. Feng and H. Xu, "Deep reinforecement learning based optimal defense for cyber-physical system in presence of unknown cyber-attack," in *Computational Intelligence (SSCI), 2017 IEEE Symposium Series on*. IEEE, 2017, pp. 1–8.

Chapter 6
Deep Feature Learning

Abstract FL is a technique that models the behavior of data from a subset of attributes only. It also shows the correlation between detection performance and traffic model quality efficiently (Palmieri et al., Concurrency Comput Pract Exp 26(5):1113–1129, 2014). However, feature extraction and feature selection are different. Feature extraction algorithms derive new features from the original features to (i) reduce the cost of feature measurement, (ii) increase classifier efficiency, and (iii) improve classification accuracy, whereas feature selection algorithms select no more than m features from a total of M input features, where m is smaller than M. Thus, the newly generated features were merely selected from the original features without any transformation. However, their goal is to derive or select a characteristic feature vector with a lower dimensionality which is used for the classification task. One advantage of deep learning models is processing underlying data from the input which suits for FL tasks. Therefore, we discuss this critical role of deep learning in IDS as Deep Feature Extraction and Selection (D-FES) and deep learning for clustering.

6.1 Deep Feature Extraction and Selection

The recent advances in mobile technologies have resulted in IoT-enabled devices becoming more pervasive and integrated into our daily lives. The security challenges that need to be overcome mainly stem from the open nature of a wireless medium such as a Wi-Fi network. An impersonation attack is an attack in which an adversary is disguised as a legitimate party in a system or communications protocol. The connected devices are pervasive, generating high-dimensional data on a large scale, which complicates simultaneous detections. FL, however, can circumvent the potential problems that could be caused by the large-volume nature of network data. Aminanto et al. [2] presented a novel D-FES, which combines stacked feature extraction and weighted feature selection. The stacked auto-encoding is capable of providing representations that are more meaningful by reconstructing the relevant

© The Author(s), under exclusive license to Springer Nature Singapore Pte Ltd. 2018   47
K. Kim et al., *Network Intrusion Detection using Deep Learning*,
SpringerBriefs on Cyber Security Systems and Networks,
https://doi.org/10.1007/978-981-13-1444-5_6

information from its raw inputs. These representations were then combined with modified weighted feature selection inspired by an existing shallow-structured machine learner. The usefulness of the condensed set of features to reduce the bias of a machine learner model as well as the computational complexity is shown.

6.1.1   Methodology

Feature extraction and selection could be adopted from D-FES. Figure 6.1 shows the stepwise procedure of D-FES with two target classes. A preprocessing procedure, which comprises of the normalization and balancing steps, is necessary. The process is explained in Sect. 6.1.2 in detail. As illustrated in Algorithm 1, D-FES starts by constructing SAE-based feature extractor with two consecutive hidden layers to optimize the learning capability and the execution time [3]. The SAE outputs 50 extracted features, which are then combined with the 154 original features existing in the AWID dataset [4]. Weighted feature selection methods were then utilized using well-referenced machine learners including SVM, ANN, and C4.5 to construct the candidate models, namely, D-FES-SVM, D-FES-ANN, and D-FES-C4.5, respectively. SVM separates the classes using a support vector (hyperplane). Then, ANN optimizes the parameters related to hidden layers that minimize the classifying error concerning the training data, whereas C4.5 adopts a hierarchical decision scheme such as a tree to distinguish each feature [5]. The final step of the detection task involves learning an ANN classifier with 12–22 trained features only.

The supervised feature selection block in Fig. 6.1 consists of three different feature selection techniques. These techniques are similar in that they consider their resulting weights to select the subset of essential features.

ANN is used as one of the weighted feature selection methods. The ANN was trained with two target classes only (normal and impersonation attack classes).

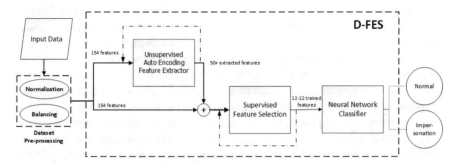

Fig. 6.1 Stepwise procedure of D-FES with two target classes: normal and impersonation attack

---

Algorithm 1 Pseudocode of D-FES

 1: **procedure** D-FES
 2:     **function** DATASET PRE-PROCESSING(Raw Dataset)
 3:         **function** (Dataset Normalization)Raw Dataset
 4:             **return** $Normalized Dataset$
 5:         **end function**
 6:         **function** (Dataset Balancing)Normalized Dataset
 7:             **return** $Balanced Dataset$
 8:         **end function**
 9:         **return** $Input Dataset$
10:     **end function**
11:     **function** DEEP ABSTRACTION($Input Dataset$)
12:         **for** i=1 to h **do**   ▷ h=2; number of hidden layers
13:             **for** each data instance **do**
14:                 Compute y_i (Eq. (3.7))
15:                 Compute z_i (Eq. (3.8))
16:                 Minimize E_i (Eq. (3.12))
17:                 $\theta_i = \{W_i, V_i, b_{f_i}, b_{g_i}\}$
18:             **end for**
19:             $W \leftarrow W_2$ ▷ 2nd layer, 50 extracted features
20:         **end for**
21:         $Input Features \leftarrow W + Input Dataset$
22:         **return** $Input Features$
23:     **end function**
24:     **function** FEATURE SELECTION($Input Features$)
25:         **switch** D-FES **do**
26:             **case** D-FES-ANN($Input Features$)
27:                 **return** $Selected Features$
28:             **case** D-FES-SVM($Input Features$)
29:                 **return** $Selected Features$
30:             **case** D-FES-C4.5($Input Features$)
31:                 **return** $Selected Features$
32:     **end function**
33:     **procedure** CLASSIFICATION($Selected Features$)
34:         Training ANN
35:         Minimize E (Eq. (4.1))
36:     **end procedure**
37: **end procedure**

Figure 6.2 shows an ANN network with one hidden layer only where b_1 and b_2 represent the bias values for the corresponding hidden and output layer, respectively.

To select the essential features, the weight values between the first two layers were considered. The weight represents the contribution from the input features to the first hidden layer. A w_{ij} value close to zero means that the corresponding input feature x_j is meaningless for further propagation, thus having one hidden layer is sufficient for this particular task. The important value of each input feature is shown in Eq. (6.1).

Fig. 6.2 ANN network with one hidden layer only

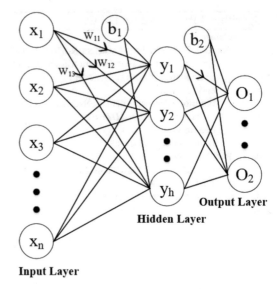

Input Layer

Algorithm 2 D-FES-ANN function

 1: **function** D-FES-ANN(*Input Features*)
 2:      Training ANN
 3:      w_{ij}
 4:      **for** each input feature **do**
 5:          Compute V_j (Eq. (6.1))
 6:      **end for**
 7:      Sort descending
 8:      *Selected Features* $\leftarrow V_j > threshold$
 9:      **return** *Selected Features*
10: **end function**

$$V_j = \sum_{i=1}^{h} |w_{ij}|, \qquad (6.1)$$

where h is the number of neurons in the first hidden layer. As described in Algorithm 2, the feature selection process involves selecting the features of which the V_j values are higher than the threshold value after the input features were sorted according to their V_j values in descending order.

Following the weighted feature selection, ANN is also used as a classifier. When learning with ANN, a minimum global error function was executed. It has two learning approaches, supervised and unsupervised. This study uses a supervised approach since knowing the class label may increase the classifier performance [6]. Also, a scaled conjugate gradient optimizer, which is suitable for a large scale problem, is used [7].

SVM-RFE is used in D-FES-SVM by using the linear case [8] described in Algorithm 3. The inputs are training instances and class labels. First, a feature

---

Algorithm 3 D-FES-SVM function

1: **function** D-FES-SVM($Input Features$)
2:      Training SVM
3:      Compute w (Eq. (3.2))
4:      Compute the ranking criteria
5:      $c_i = w_i{}^2$
6:      Find the smallest ranking criterion
7:      $f = argmin(c)$
8:      Update feature ranked list
9:      $r = [s(f), r]$
10:     Eliminate the smallest ranking criterion
11:     $s = s(1 : f - 1, f + 1 : length(s))$
12:     $Selected Features \leftarrow s$
13:     **return** $Selected Features$
14: **end function**

---

ranked list was initialized that is filled by a subset of important features that are used for selecting training instances. The scheme continues by training the classifier and computes the weight vector of the dimension length. After the value of the weight vector was obtained, it computes the ranking criteria and finds the feature with the smallest ranking criterion. Using that feature, the feature ranking list was updated, and the feature with the smallest ranking criterion was eliminated. A feature ranked list was finally created as its output.

The last feature selection method is using DT C4.5. The feature selection process begins by selecting the top-three level nodes as explained in Algorithm 4. It then removes the equal nodes and updates the list of selected features.

Algorithm 4 D-FES-C4.5 function

1: **function** D-FES-C4.5($Input Features$)
2:      Training C4.5
3:      $Selected Features \leftarrow$ top-three level nodes
4:      **for** i=1 to n **do**   ▷ n = size of $Selected Features$
5:          **for** j=1 to n **do**
6:              **if** $Selected Features[i]=Selected Features[j]$ **then**   Remove $Selected Features[j]$
7:              **end if**
8:          **end for**
9:      **end for**
10:     **return** $Selected Features$
11: **end function**

6.1.2   Evaluation

A set of experiments was conducted to evaluate the performance of the proposed D-FES method in Wi-Fi impersonation detection. Choosing a proper dataset is an important step in the IDS research field [9]. The AWID dataset [4] which comprises of Wi-Fi network data collected from real network environments is used in this study. Fair model comparison and evaluation were achieved by performing the experiments on the same testing sets as in [4]. The method was implemented using MATLAB R2016a and Java code extracted and modified from WEKA packages [10] running on an Intel Xeon E-3-1230v3 CPU @3.30 GHz with 32 GB RAM.

6.1.2.1   Dataset Preprocessing

The data contained in the AWID dataset are diverse in value, discrete, continuous, and symbolic, with a flexible value range. These data characteristics could make it difficult for the classifiers to learn the underlying patterns correctly [11]. The preprocessing phase thus includes mapping symbolic-valued attributes to numeric values, according to the normalization steps and dataset-balancing process described in Algorithm 5. The target classes were mapped to one of these integer-valued classes: 1 for normal instances, 2 for an impersonation, 3 for flooding, and 4 for an injection attack. Meanwhile, symbolic attributes such as a receiver, destination, transmitter, and source address were mapped to integer values with a minimum value of 1 and a maximum value, which is the number of all symbols. Some dataset attributes such as the WEP Initialization Vector (IV) and Integrity Check Value (ICV) were hexadecimal data, which need to be transformed into integer values as well. The continuous data such as the timestamps were also left for the normalization step. Some of the attributes have question marks, ?, to indicate

---

Algorithm 5 Dataset pre-processing function

```
 1: function DATASET PRE-PROCESSING(Raw Dataset)
 2:     function DATASET NORMALIZATION(Raw Dataset)
 3:         for each data instance do
 4:             cast into integer value
 5:             normalize (Eq. (6.2))
 6:             Normalized Dataset
 7:         end for
 8:     end function
 9:     function DATASET BALANCING(Normalized Dataset)
10:         Pick 10% of normal instances randomly
11:         Balanced Dataset
12:     end function
13:     Input Dataset ← Balanced Dataset
14:     return Input Dataset
15: end function
```

Table 6.1 Distribution of each class for both balanced and unbalanced dataset

Class		Training	Test
Normal	Unbalanced	1,633,190	530,785
	Balanced	163,319	53,078
Attack	Impersonation	48,522	20,079
	Flooding	48,484	8,097
	Injection	65,379	16,682
	Total	162,385	44,858

AWID dataset mimics the natural unbalanced network distribution between normal and attack instances. "Balanced" means to make equal distribution between the number of normal instances (163,319) and thereof total attack instances (162,385). 15% of training data were withdrawn for validation data.

unavailable values. One alternative was selected in which the question mark was assigned to a constant zero value [12]. After all, data were transformed into numerical values; attribute normalization is needed [13]. Data normalization is a process; hence, all value ranges of each attribute were equal. The mean range method [14] was adopted in which each data item is linearly normalized between zero and one to avoid the undue influence of different scales [12]. Equation (6.2) shows the normalizing formula:

$$z_i = \frac{x_i - min(x)}{max(x) - min(x)},\tag{6.2}$$

where z_i denotes the normalized value, x_i refers to the corresponding attribute value, and $min(x)$ and $max(x)$ are the minimum and maximum values of the attribute, respectively.

The reduced "CLS" data are a good representation of a real network, in which normal instances significantly outnumber attack instances. The ratio between the normal and attack instances is 10:1 for both unbalanced training and the test dataset as shown in Table 6.1. This property might be biased to the training model and affect the model performance [15, 16]. To alleviate this, the dataset was balanced by selecting 10% of the normal instances randomly. However, a specific value was set as the seed of the random number generator for reproducibility purposes. The ratio between normal and attack instances became 1:1, which is an appropriate proportion for the training phase [16]. D-FES was trained using the balanced dataset and then verified on the unbalanced dataset.

6.1.2.2 Experimental Result

The proposed D-FES was evaluated on a set of experiments. First, different architectures of the feature extractor were implemented and verified, SAE. Second, two feature selection approaches: filter-based and wrapper-based methods were

Table 6.2 The evaluation of SAE's schemes

SAE Scheme	DR (%)	FAR (%)	Acc (%)	F_1 (%)
Imbalance_40 (154:40:10:4)	64.65	**1.03**	96.30	73.13
Imbalance_100 (154:100:50:4)	85.30	1.98	**97.03**	81.75
Balance_40 (154:40:10:4)	72.58	18.87	77.21	74.48
Balance_100 (154:100:50:4)	**95.35**	18.90	87.63	**87.59**

verified. Last, the usefulness and the utility of D-FES was validated on a realistic unbalanced test dataset.

The SAE architectures were varied to optimize the SAEs implementation with two hidden layers. The features generated from the first encoder layer were employed as the training data in the second encoder layer. Meanwhile, the size of each hidden layer was decreased accordingly such that the encoder in the second encoder layer learns an even smaller representation of the input data. The regression layer with the softmax activation function was then implemented in the final step. The four schemes were examined to determine the SAE learning characteristics. The first scheme, Imbalance_40, has two hidden layers with 40 and 10 hidden neurons in each layer. The second scheme, Imbalance_100, also has two hidden layers; however, it employs 100 and 50 hidden neurons in each layer. Although there is no strict rule for determining the number of hidden neurons, we consider a common rule of thumb [17], which ranges from 70% to 90% from inputs. The third and fourth schemes, named Balance_40 and Balance_100, have the same hidden layer architecture with the first and second schemes, respectively; however, in this case, the balanced dataset was used, because the common assumption is that a classifier model built by a highly unbalanced data distribution performs poorly on minority class detection [18]. For testing purposes, all four classes contained in the AWID dataset was used.

Table 6.2 shows the evaluation of the SAE schemes. Each model uses either balanced or unbalanced data for the SAE algorithm with the following parameters: input features, number of features in 1st hidden layer, number of features in 2nd hidden layer, and target classes. The SAE architectures with 100 hidden neurons have higher DR than those with 40 hidden neurons. On the other hand, the SAE architectures with 40 hidden neurons have lower FAR than those with 100 hidden neurons. To draw a proper conclusion, other performance metrics that consider whole classes are needed as the DR checks for the attack class only and the FAR measures for the normal class only. The Acc metric would be affected by the distribution of data, for which different balanced and unbalanced distributions may result in an incorrect conclusion. If we consider the Acc metric only as in Fig. 6.3, we may incorrectly select the Imbalance_100 with 97.03% accuracy, whereas the Balance_100 only achieved 87.63% accuracy. In fact, the Imbalance_100 achieved the highest accuracy rate because of the unbalanced proportion of normal class to attack class. The best performance was obtained by checking F_1 score, for which the Balance_100 has achieved the highest F_1 score among all schemes with 87.59%. Therefore, the SAE architecture with 154:100:50:4 topology was chosen.

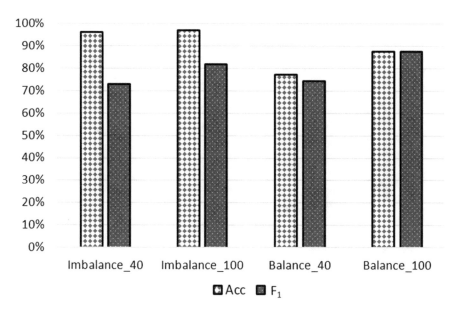

Fig. 6.3 Evaluation of SAE's scheme on Acc and F_1 score. The red bar represents F_1 score while the blue bar represents Acc rate

To show the effectiveness of D-FES, the following feature selection methods were compared:

- CfsSubsetEval [19] (CFS) considers the predictive ability of each feature individually and the degree of redundancy between them, to evaluate the importance of a subset of features. This approach selects subsets of features that are highly correlated with the class while having low intercorrelation.
- Correlation (Corr) measures the correlation between the feature and the class to evaluate the importance of a subset of features.
- The weight from a trained ANN model mimics the importance of the correspondence input. By selecting the important features only, the training process becomes lighter and faster than before [20].
- SVM measures the importance of each feature based on the weight of the SVM classification results.
- C4.5 is one of the DT approaches. It can select a subset of features that are not highly correlated. Correlated features should be in the same split; hence, features that belong to different splits are not highly correlated [21].

A filter-based method usually measures the correlation and redundancy of each attribute without executing a learning algorithm. Therefore, the filter-based method is typically lightweight and fast. On the other hand, the wrapper-based method examines the results of any learning algorithm that outputs a subset of features [22]. CFS and Corr belong to the filter-based techniques, whereas ANN, SVM, and C4.5 are wrapper-based methods.

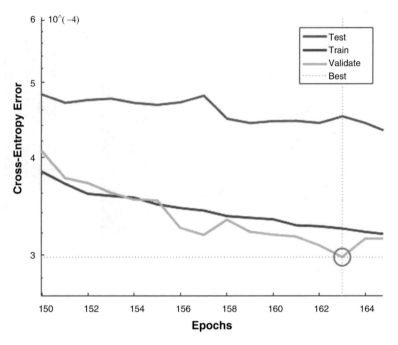

Fig. 6.4 Cross-entropy error of ANN. The best validation performance was achieved at the epoch of 163

A subset of features was selected using the wrapper-based method for considering each feature weight. For ANN, a threshold weight value was defined, and if the weight of a feature is higher than the threshold, then the feature is selected. The SVM attribute selection function ranks the features based on their weight values. The subset of features with a higher weight value than the predefined threshold value was then selected. Similarly, C4.5 produces a deep binary tree. The features were chosen that belong to the top-three levels in the tree. CFS produces a fixed number of selected features and Corr provides a correlated feature list.

During ANN training for both feature selection and classification, the trained model was optimized using a separate validation dataset; that is, the dataset was separated into three parts, training data, validation data, and testing data in the following proportions: 70%, 15%, and 15%, respectively. The training data were used as input into the ANN during training, and the weights of neurons were adjusted during the training according to its classification error. The validation data were used to measure model generalization providing useful information on when to terminate the training process. The testing data was used for an independent measure of the model performance after training. The model is said to be optimized when it reaches the smallest average square error on the validation dataset. Figure 6.4 shows an example of ANN performance concerning the cross-entropy error during ANN training. At the epoch of 163, the cross-entropy error, a logarithmic-based error

Table 6.3 Feature set comparisons between feature selection and D-FES

Method	Selected Features	D-FES
CFS	5, 38, 70, 71, 154	38, 71, 154, 197
Corr	47, 50, 51, 67, 68, 71, 73, 82	71, 155, 156, 159, 161, 165, 166, 179, 181, 191, 193, 197
ANN	4, 7, 38, 77, 82, 94, 107, 118	4, 7, 38, 67, 73, 82, 94, 107, 108, 111, 112, 122, 138, 140, 142, 154, 161, 166, 192, 193, 201, 204
SVM	47, 64, 82, 94, 107, 108, 122, 154	4, 7, 47, 64, 68, 70, 73, 78, 82, 90, 94, 98, 107, 108, 111, 112, 122, 130, 141, 154, 159
C4.5	11, 38, 61, 66, 68, 71, 76, 77, 107, 119, 140	61, 76, 77, 82, 107, 108, 109, 111, 112, 119, 158, 160

measurement comparing the output values and desired values, starts increasing, meaning that at the epoch of 163, the model was optimized. Although the training data output decreasing error values after the epoch point of 163, the performance of the model no longer continues to improve, as the decreasing cross-entropy error may indicate the possibility of overfitting.

Table 6.3 contains all the feature lists selected from the various feature selection methods. Some features were essential for detecting an impersonation attack. These are the 4th and the 7th, which were selected by the ANN and SVM, and the 71st, which is selected by CFS and Corr. The characteristics of the selected features are shown in Fig. 6.5a–b. The blue line indicates normal instances, and at the same time, the red line depicts the characteristics of an impersonation attack. Normal and attack instances can be distinguished based on the attribute value of data instances. For example, once a data instance has an attribute value of 0.33 in the 166th feature, the data instance has a high probability of being classified as an attack. This could be applied to the 38th and other features as well.

Table 6.4 lists the performance of each algorithm on the selected feature set only. SVM achieved the highest DR (99.86%) and Mcc (99.07%). However, it requires CPU time of 10,789s to build a model, the longest time among the models observed. As expected, the filter-based methods (CFS and Corr) built their models quickly; however, they attained the lowest Mcc for CFS (89.67%).

Table 6.5 compares the performances of the candidate models on the feature sets that were produced by D-FES. SVM again achieved the highest DR (99.92%) and Mcc (99.92%). It also achieved the highest FAR with a value of only 0.01%. Similarly, the lowest Mcc was achieved by Corr (95.05%). This concludes that wrapper-based feature selections outperform filter-based feature selections. As SVM showed the best performance, the properties of selected features by SVM are described in Table 6.6.

The following patterns were observed from Tables 6.4 and 6.5: Only two out of five methods (Corr and C4.5) showed lower FAR without D-FES, which is expected to minimize the FAR value of the proposed IDS. This phenomenon might exist because the original and extracted features were not correlated because Corr and C4.5 measure the correlation between each feature. Filter-based feature selection

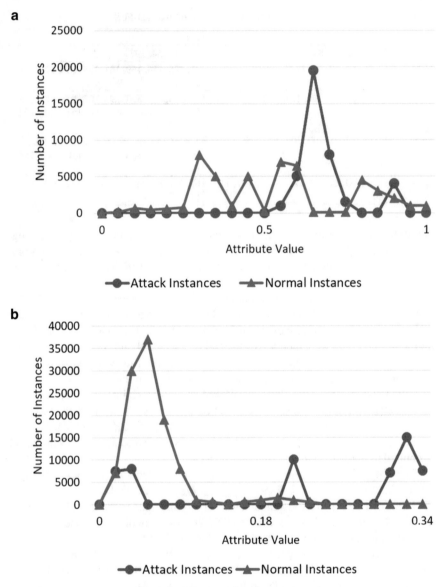

Fig. 6.5 Characteristics of (**a**) 38th and (**b**) 166th features. The blue line represents normal instances while the red line represents attack instances

methods require much shorter CPU time compared to the CPU time taken by D-FES. However, D-FES improves the filter-based feature selections performance significantly.

Similar patterns were captured by Fig. 6.6a–c, which depict the performance of different models in terms of Acc, $Precision$, and F_1 score, respectively. D-FES-SVM achieved the highest Acc, $Precision$, and F_1 score of 99.97%, 99.96%,

Table 6.4   Model comparisons on selected features

Model	DR (%)	FAR (%)	Acc (%)	F_1 (%)	Mcc (%)	TBM (s)
CFS	94.85	3.31	96.27	92.04	89.67	80
Corr	92.08	0.39	97.88	95.22	93.96	**2**
ANN	99.79	0.47	97.88	99.10	98.84	150
SVM	**99.86**	0.39	**99.67**	99.28	99.07	10,789
C4.5	99.43	**0.23**	99.61	**99.33**	**99.13**	1,294

Table 6.5   Model comparisons on D-FES feature set

Model	DR (%)	FAR (%)	Acc (%)	F_1 (%)	Mcc (%)	TBM (s)
CFS	96.34	0.46	98.80	97.37	96.61	1,343
Corr	95.91	1.04	98.26	96.17	95.05	**1,264**
ANN	99.88	0.02	99.95	99.90	99.87	1,444
SVM	**99.92**	**0.01**	**99.97**	**99.94**	**99.92**	12,073
C4.5	99.55	0.38	99.60	99.12	98.86	2,595

Table 6.6   Feature set selected by D-FES-SVM

Index	Feature Name	Description
47	radiotap.datarate	Data rate (Mb/s)
64	wlan.fc.type_subtype	Type or Subtype
82	wlan.seq	Sequence number
94	wlan_mgt.fixed.capabilities.preamble	Short Preamble
107	wlan_mgt.fixed.timestamp	Timestamp
108	wlan_mgt.fixed.beacon	Beacon Interval
122	wlan_mgt.tim.dtim_period	DTIM period
154	data.len	Length

and 99.94%, respectively. By D-FES, all methods achieve *Precision* of more than 96%, which shows that D-FES can reduce the number of incorrect classification of normal instances as an attack. Also, D-FES improves the *Acc* of filter-based feature selections significantly. Except for the C4.5, all feature selection methods were improved both the *Acc* and F_1 score by using D-FES. We compare D-FES-SVM as the highest F1 score, and randomly selected features with respect to the number of features involved during training as depicted in Fig. 6.7. D-FES-SVM takes longer time than the random method and it increases DR significantly. However, the random method cannot even classify a single impersonation attack. This makes the proposed D-FES a good candidate for an intrusion detector.

6.2   Deep Learning for Clustering

IDS has been becoming a vital measure in any networks, especially Wi-Fi networks. Wi-Fi networks growth is undeniable due to a vast amount of tiny devices connected via Wi-Fi networks. Regrettably, adversaries may take advantage by launching an impersonation attack, a typical wireless network attack. Any IDS

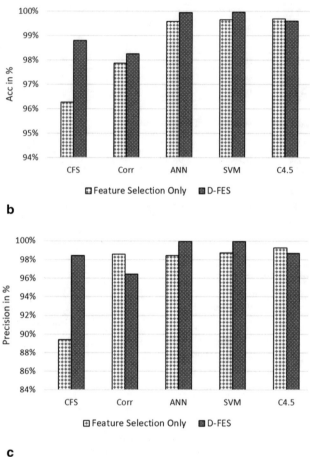

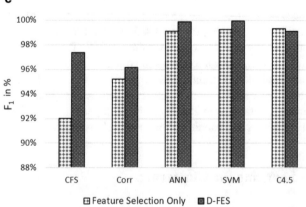

Fig. 6.6 Model performance comparisons in terms of (**a**) Acc, (**b**) $Precision$, and (**c**) F_1 score. The blue bar represents performances by feature selection only while the red bar represents performances by D-FES

Fig. 6.7 Model performance comparisons between D-FES and random method in terms of (**a**) DR, (**b**) FAR, (**c**) FNR, and (**d**) TT

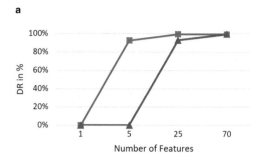

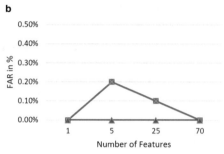

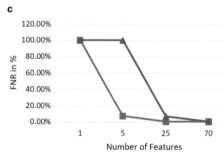

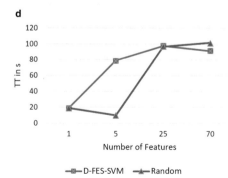

usually depends on classification capabilities of machine learning, which supervised learning approaches give the best performance to distinguish benign and malicious data. However, due to massive traffic, it is difficult to collect labeled data in Wi-Fi networks. Therefore, Aminanto and Kim [23] proposed a novel fully unsupervised method which can detect attacks without prior information on data label. The method is equipped with an unsupervised SAE for extracting features and a k-means clustering algorithm for clustering task.

6.2.1 Methodology

In this section, the novel fully unsupervised deep learning-based IDS for detecting impersonation attacks is explained. There are two main tasks, feature extraction, and clustering tasks. Figure 6.8 shows the scheme which contains two main functions in cascade. A real Wi-Fi networks-trace, AWID dataset [4] is used, which contains 154 original features. Before the scheme starts, normalizing and balancing process should be done to achieve best training performance. Algorithm 6 explains the procedure of the scheme in detail.

The scheme starts with two cascading encoders, and the output features from the second layer are then forwarded to the clustering algorithm. The first encoder has 100 neurons as the first hidden layer, while the second encoder comes with 50 neurons only. A standard rule for choosing the number of neurons in a hidden layer is using 70% to 90% of the previous layer. In this paper, $k = 2$ was defined since they considered two classes only. The scheme ends by two clusters formed by k-means clustering algorithm. These clusters represent benign and malicious data.

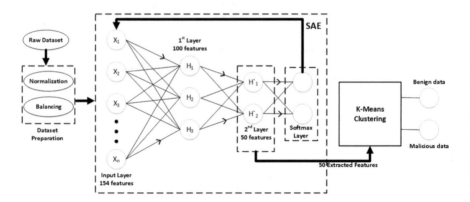

Fig. 6.8 Our proposed scheme contains feature extraction and clustering tasks

---

Algorithm 6 Pseudocode of fully unsupervised deep learning

1: **procedure** START
2:     **function** DATASET PREPARATION(Raw Dataset)
3:         **for** each data instance **do**
4:             Convert into integer value
5:             Normalization $z_i = \frac{x_i - min(x)}{max(x) - min(x)}$
6:         **end for**
7:         Balance the normalized dataset
8:         **return** $Input\,Dataset$
9:     **end function**
10:     **function** SAE($Input\,Dataset$)
11:         **for** i=1 to h **do**   ▷ h=2; number of hidden layers
12:             **for** each data instance **do**
13:                 Compute $H = s_f\left(WX + b_f\right)$
14:                 Compute $X' = s_g\left(VH + b_g\right)$
15:                 Minimize $E = \frac{1}{N}\sum_{n=1}^{N}\sum_{k=1}^{K}\left(X'_{kn} - X_{kn}\right)^2 + \lambda \cdot \Omega_{weights} + \beta \cdot \Omega_{sparsity}$
16:                 $\theta_i = \{W_i, V_i, b_{f_i}, b_{g_i}\}$
17:             **end for**
18:             $Input\,Features \leftarrow W_2$ ▷ 2nd layer, 50 extracted features
19:         **end for**
20:         **return** $Input\,Features$
21:     **end function**
22:     Initialize clusters and k=2   ▷ two clusters: benign and malicious
23:     **function** k-MEANS CLUSTERING($Input\,Features$)
24:         **return** $Clusters$
25:     **end function**
26:     Plot confusion between $Clusters$ and target classes
27: **end procedure**

6.2.2 Evaluation

There are two hidden layers in the SAE network with 100 and 50 neurons accordingly. The encoder in the second layer fed with features formed by the first layer of the encoder. The softmax activation function was implemented in the final stage of the SAE to optimize the SAE training. The 50 features extracted from the SAE were then forwarded to k-means clustering algorithm as input. Random initialization was used for k-means clustering algorithm. However, a particular value must be defined as a random number seed for reproducibility purpose. Clustering results were compared from three inputs: original data, features from the first hidden layer of the SAE, and features from the second hidden layer of the SAE as shown in Table 6.7.

It was observed that the limitation of a traditional k-means algorithm, which is unable to cluster complex and high-dimensional data of AWID dataset, as expressed by 55.93% of accuracy only. Although 100 features coming from the 1st hidden layer achieved 100% of DR, the false alarm rate was still unacceptable with 57.48%.

Table 6.7 The evaluation of our proposed scheme

Input	$DR(\%)$	$FAR(\%)$	$Acc(\%)$	$Precision(\%)$	$F_1(\%)$
Original data	100.00	57.17	55.93	34.20	50.97
1st hidden layer	100.00	57.48	55.68	34.08	50.83
2nd hidden layer	92.18	4.40	94.81	86.15	89.06

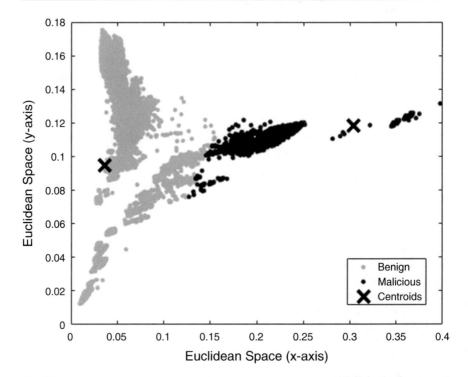

Fig. 6.9 Cluster assignment result in Euclidean space by our proposed scheme

The k-means algorithm fed by 50 features from the 2nd hidden layer achieved the best performance among all as shown by the highest F_1 score (89.06%) and Acc (94.81%), also the lowest FAR (4.40%). Despite a bit lower DR, the scheme improved the traditional k-means algorithm in overall by almost twice F_1 score and accuracy.

Figure 6.9 shows cluster assignment result in Euclidean space using this scheme. Black dots represent attack instances, while gray dots represent benign instances. The location of cluster centroid for each cluster is expressed by X mark.

The performance of the scheme also compared against two previous related work by Kolias et al.[4] and Aminanto and Kim [24] as shown in Table 6.8. The scheme can classify impersonation attack instances with a DR of 92.18% while maintaining low FAR, 4.40%. Kolias et al. [4] tested various classification algorithms such as Random Tree, Random Forest, J48, Naive Bayes, etc., on AWID dataset. Among

Table 6.8 Comparison with previous work

Method	$DR(\%)$	$FAR(\%)$	$Acc(\%)$	$Precision(\%)$	$F_1(\%)$
Kolias et al.[4]	22.01	0.02	97.14	97.57	35.92
Aminanto and Kim [24]	65.18	0.14	98.59	94.53	77.16
Our proposed scheme [23]	92.18	4.40	94.81	86.15	89.06

Table 6.9 IDSs leveraging SAE

Publication	Role of SAE	Combined with
AK16a [24]	Classifier	ANN
AK16b [28]	Feature Extractor	Softmax Regression
AK17 [23]	Clustering	K-means Clustering
ACTYK17 [2]	Feature Extractor	SVM, DT, ANN

all methods, Naive Bayes algorithm showed the best performance by correctly classifying 4,419 out of 20,079 impersonation instances. It achieved approximately 22% DR only, which is unsatisfactory. Aminanto and Kim [25] proposed another impersonation detector by combining ANN with SAE, which successfully improved the IDS model for impersonation attack detection task by achieving a DR of 65.18% and a FAR of 0.14%. In this study, SAE was leveraged for assisting traditional k-means clustering with extracted features. Although the scheme resulted in a high false alarm rate, which leads to a severe impact of IDS [26], however, this false alarm rate value about 4% can be acceptable since the fully unsupervised approach was used. The parameters can be adjusted and cut the FAR down, but less FAR or high DR remains a trade-off for practices and needs to be investigated in the future. It is observed that the advantage of SAE is abstracting a complex and high-dimensional data to assist traditional clustering algorithm which is shown by reliable DR and F_1 score achieved by the scheme.

6.3 Comparison

The goal of deep learning method is learning feature hierarchies from the lower level to higher level features [27]. The technique can learn features independently at multiple levels of abstraction and thus discover complicated functions mapping between the input and the output directly from raw data without depending on customized features by the experts. In higher-level abstractions, humans often have no idea to see the relation and connection from the raw sensory input. Therefore, the ability to learn sophisticated features, also called as feature extraction, must be highlighted as the amount of data increases sharply [27]. SAE is one good instance of feature extractors. Therefore, several previous works which implement SAE as the feature extractor and other roles in the IDS module are discussed as shown in Table 6.9.

Table 6.10 Comparison on impersonation detection

Method	DR (%)	False Alarm Rate (%)
AK16a [24]	65.178	0.143
AK16b [28]	92.674	2.500
AK17 [23]	92.180	4.400
ACTYK17 [2]	99.918	0.012
KKSG15 [4]	22.008	0.021

Feature extraction by SAE can reduce the complexity of original features of the dataset. However, besides a feature extractor, SAE can also be used for classifying and clustering tasks as shown in Table 6.9. AK16b [28] used semi-supervised approach for IDS which contains feature extractor (unsupervised learning) and classifier (supervised learning). SAE was leveraged for feature extraction and regression layer with softmax activation function for the classifier. SAE as feature extractor also used in ACTYK17 [2], but ANN, DT, and SVM were leveraged as a feature selection. In other words, it combines stacked feature extraction and weighted feature selections. By experiment results [2], D-FES improved the feature learning process by combining stacked feature extraction with weighted feature selection. The feature extraction of SAE is capable of transforming the original features into a more meaningful representation by reconstructing its input and providing a way to check that the relevant information in the data has been captured. SAE can be efficiently used for unsupervised learning on a complex dataset.

Unlike two previous approaches, AK16a [24] and AK17 [23] used SAE for other roles than a feature extractor, namely, classifying and clustering methods, respectively. ANN was adopted as a feature selection since the weight from trained models mimics the significance of the corresponding input [24]. By selecting the important features only, the training process becomes lighter and faster than before. AK16a [24] exploited SAE as a classifier since this employs consecutive layers of processing stages in hierarchical manners for pattern classification and feature or representation learning. On the other hand, AK17 [23] proposed a novel fully unsupervised method which can detect attacks without prior information on data label. The scheme is equipped with an unsupervised SAE for extracting features and a k-means clustering algorithm for clustering task.

Kolias et al. [4] tested many existing machine-learning models on the dataset in a heuristic manner. The lowest DR is observed particularly on impersonation attack reaching an accuracy of 22% only. Therefore, improving impersonation detection is challenging, and hence the comparison of previous approaches on impersonation detection are summarized in Table 6.10. DR refers to the number of attacks detected divided by the total number of attack instances in the test dataset, while FAR is the number of normal instances classified as an attack divided by the total number of normal instances in the test dataset.

From Table 6.10, it is observed that SAE can improve the performance of IDS compared to KKSG15 [4]. It is verified that SAE achieved high-level abstraction of complex and huge Wi-Fi network data. The SAE's model free properties and learnability on complex and large-scale data fit into the open nature of Wi-Fi networks. Among all IDSs, the one using SAE as a classifier achieved the lowest impersonation attack DR with 65.178% only. It shows that SAE can be a classifier but not excellent as the original role of SAE is a feature extractor. The usability of SAE as a feature extractor is validated by AK16b [28] and ACTYK17 [2] which achieved the highest DR. Even more, by a combination of SAE extractor and weighted selection [2], the best performance of DR and FAR among other was achieved. Besides that, an interesting fact is that SAE can assist k-means clustering algorithm to achieve better performance with DR of 92.180% [23]. However, it is required to analyze further to reduce the FAR since higher FAR will be undesirable for practical IDS's.

References

1. F. Palmieri, U. Fiore, and A. Castiglione, "A distributed approach to network anomaly detection based on independent component analysis," *Concurrency and Computation: Practice and Experience*, vol. 26, no. 5, pp. 1113–1129, 2014.
2. M. E. Aminanto, R. Choi, H. C. Tanuwidjaja, P. D. Yoo, and K. Kim, "Deep abstraction and weighted feature selection for Wi-Fi impersonation detection," *IEEE Transactions on Information Forensics and Security*, vol. 13, no. 3, pp. 621–636, 2018.
3. Q. Xu, C. Zhang, L. Zhang, and Y. Song, "The learning effect of different hidden layers stacked autoencoder," in *Proc. Int. Con. Intelligent Human-Machine Systems and Cybernetics (IHMSC), Zhejiang, China*, vol. 02. IEEE, Aug 2016, pp. 148–151.
4. C. Kolias, G. Kambourakis, A. Stavrou, and S. Gritzalis, "Intrusion detection in 802.11 networks: empirical evaluation of threats and a public dataset," *IEEE Commun. Surveys Tuts.*, vol. 18, no. 1, pp. 184–208, 2015.
5. H. Shafri and F. Ramle, "A comparison of support vector machine and decision tree classifications using satellite data of langkawi island," *Information Technology Journal*, vol. 8, no. 1, pp. 64–70, 2009.
6. L. Guerra, L. M. McGarry, V. Robles, C. Bielza, P. Larrañaga, and R. Yuste, "Comparison between supervised and unsupervised classifications of neuronal cell types: a case study," *Developmental neurobiology*, vol. 71, no. 1, pp. 71–82, 2011.
7. M. F. Møller, "A scaled conjugate gradient algorithm for fast supervised learning," *Neural Networks*, vol. 6, no. 4, pp. 525–533, 1993.
8. I. Guyon, J. Weston, S. Barnhill, and V. Vapnik, "Gene selection for cancer classification using support vector machines," *Machine Learning*, vol. 46, no. 1–3, pp. 389–422, 2002.
9. A. Özgür and H. Erdem, "A review of KDD99 dataset usage in intrusion detection and machine learning between 2010 and 2015," *PeerJ PrePrints*, vol. 4, p. e1954v1, 2016.
10. M. Hall, E. Frank, G. Holmes, B. Pfahringer, P. Reutemann, and I. H. Witten, "The weka data mining software: an update," *ACM SIGKDD Explorations Newsletter*, vol. 11, no. 1, pp. 10–18, 2009.
11. M. Sabhnani and G. Serpen, "Application of machine learning algorithms to KDD intrusion detection dataset within misuse detection context." in *Proc. Int. Conf. Machine Learning; Models, Technologies and Applications (MLMTA), Lax Vegas, USA*, 2003, pp. 209–215.
12. D. T. Larose, *Discovering knowledge in data: an introduction to data mining*. John Wiley & Sons, 2014.
13. H. Bostani and M. Sheikhan, "Modification of supervised OPF-based intrusion detection systems using unsupervised learning and social network concept," *Pattern Recognition*, vol. 62, pp. 56–72, 2017.

14. W. Wang, X. Zhang, S. Gombault, and S. J. Knapskog, "Attribute normalization in network intrusion detection," in *Proc. Int. Symp. Pervasive Systems, Algorithms, and Networks (ISPAN), Kaohsiung, Taiwan*. IEEE, Dec. 2009, pp. 448–453.

15. N. Y. Almusallam, Z. Tari, P. Bertok, and A. Y. Zomaya, "Dimensionality reduction for intrusion detection systems in multi-data streams – a review and proposal of unsupervised feature selection scheme," *Emergent Computation*, vol. 24, pp. 467–487, 2017. [Online]. Available: https://doi.org/10.1007/978-3-319-46376-6_22

16. Q. Wei and R. L. Dunbrack Jr, "The role of balanced training and testing data sets for binary classifiers in bioinformatics," *Public Library of Science (PLoS) one*, vol. 8, no. 7, pp. 1–12, 2013.

17. Z. Boger and H. Guterman, "Knowledge extraction from artificial neural network models," in *Proc. Int. Conf. Systems, Man, and Cybernetics, Orlando, USA*, vol. 4. IEEE, 1997, pp. 3030–3035.

18. G. M. Weiss and F. Provost, *The effect of class distribution on classifier learning: an empirical study*. Dept. of Computer Science, Rutgers University, New Jersey, Tech. Rep. ML-TR-44, 2001.

19. M. A. Hall and L. A. Smith, "Practical feature subset selection for machine learning," in *Proc. Australasian Computer Science Conference (ACSC), Perth, Australia*. Springer, 1998, pp. 181–191.

20. Z. Wang, "The applications of deep learning on traffic identification," in *Conf. BlackHat, Las Vegas, USA*. UBM, 2015.

21. C. A. Ratanamahatana and D. Gunopulos, "Scaling up the naive Bayesian classifier: Using decision trees for feature selection," in *Workshop on Data Cleaning and Preprocessing (DCAP) at IEEE Int. Conf. Data Mining (ICDM), Maebashi, Japan*. IEEE, Dec 2002.

22. R. Kohavi and G. H. John, "Wrappers for feature subset selection," *Artificial Intelligence*, vol. 97, no. 1, pp. 273–324, 1997.

23. M. E. Aminanto and K. Kim, "Improving detection of Wi-Fi impersonation by fully unsupervised deep learning," *Information Security Applications: 18th International Workshop, WISA 2017*, 2017.

24. ——, "Detecting impersonation attack in Wi-Fi networks using deep learning approach," *Information Security Applications: 17th International Workshop, WISA 2016*, 2016.

25. ——, "Detecting impersonation attack in Wi-Fi networks using deep learning approach," in *Proc. Workshop of Information Security Applications (WISA), Jeju Island, Korea*. Springer, 2016, pp. 136–147.

26. R. Sommer and V. Paxson, "Outside the closed world: On using machine learning for network intrusion detection," in *Proc. Symp. Security and Privacy, Berkeley, California*. IEEE, 2010, pp. 305–316.

27. Y. Bengio *et al.*, "Learning deep architectures for ai," *Foundations and trends® in Machine Learning*, vol. 2, no. 1, pp. 1–127, 2009.

28. M. E. Aminanto and K. Kim, "Detecting active attacks in Wi-Fi network by semi-supervised deep learning," *Conference on Information Security and Cryptography 2017 Winter*, 2016.

Chapter 7
Summary and Further Challenges

Abstract This last chapter concludes this monograph by providing a closing statement regarding the advantage of using deep learning models for IDS purposes and why those models can improve IDS performance. Afterward, the overview of challenges and future research directions in deep learning applications for IDS is suggested.

In summary, deep learning is a derivative of machine-learning models, where it exploits the cascaded layers of data processing stages in a hierarchical structure for UFL and pattern classification. The principle of deep learning is to process hierarchical features of the provided input data, where the higher-level features are composed of lower-level features. Furthermore, the deep learning models can integrate a feature extractor and classifier into one framework which learns feature representations from unlabeled data autonomously, and thus the security experts don't need to craft the desired features manually [1]. Essentially, deep learning methods can discover sophisticated underlying structure/feature from abstract aspects. This abstraction ability of deep learning makes it feasible to abstract benign or malicious features among the provided data [2].

The objective of deep learning modeling is to learn and output feature representation which makes those models more suitable for feature engineering. Feature engineering here includes feature/representation learning and feature selection [3]. The power of modeling the traffic behavior from the most characterizing raw input internal dynamics is of crucial importance to show the correlation between anomaly detection performance and the traffic model quality [4].

Further challenges are left for improving IDS in the future. Based on our previous work, we recommend the followings for future directions in IDS researches, but are not limited to:

1. Training load in deep learning methods is usually huge. One should combine DNN with an asynchronous multi-threaded search that executes simulations on CPU and computes policy and value networks in parallel on GPUs. Therefore,

© The Author(s), under exclusive license to Springer Nature Singapore Pte Ltd. 2018   69
K. Kim et al., *Network Intrusion Detection using Deep Learning*,
SpringerBriefs on Cyber Security Systems and Networks,
https://doi.org/10.1007/978-981-13-1444-5_7

how to apply this deep learning model in a constrained-computation device is a really challenging task. We should make it lighter to be suitable for IoT environments such as Controller Area Network (CAN) used by Unmanned Vehicle.

2. Incorporating deep learning models as a real-time classifier will be challenging. In the most previous works that leverage deep learning methods in their IDS environment, they perform the feature extraction or reduce feature dimensionalities. Even more, a complete dataset with class labels is not easy to get. However, deep learning models are still a suitable method to analyze huge data.

3. Improving unsupervised approach since huge labeled data are difficult to obtain. Therefore an IDS leveraging unsupervised approach is desirable.

4. Build an IDS that is able to detect zero-day attacks with high detection rate and low false alarm rate.

5. A comprehensive measure not only detection but also prevention is needed in the future. Therefore, building an IDS with both detection and prevention capabilities (e.g., Intrusion Prevention System (IPS)) is expected.

6. A time series analysis by using LSTM networks promises a good anomaly detector. However, again, the training workload is still high for real-time analysis. Therefore, lightweight models of this network are desirable as shown in [5].

7. CNN achieved outstanding results in many research areas, especially in image recognition fields. However, in IDS researches, not many works benefited by using CNN. We expect that by applying a proper text-to-image conversion, we may benefit full potential of CNN as already shown in image recognition researches.

References

1. Y. Wang, W.-d. Cai, and P.-c. Wei, "A deep learning approach for detecting malicious javascript code," *Security and Communication Networks*, vol. 9, no. 11, pp. 1520–1534, 2016.
2. W. Jung, S. Kim, and S. Choi, "Poster: deep learning for zero-day flash malware detection," in *36th IEEE symposium on security and privacy*, 2015.
3. P. Louvieris, N. Clewley, and X. Liu, "Effects-based feature identification for network intrusion detection," *Neurocomputing*, vol. 121, pp. 265–273, 2013.
4. F. Palmieri, U. Fiore, and A. Castiglione, "A distributed approach to network anomaly detection based on independent component analysis," *Concurrency and Computation: Practice and Experience*, vol. 26, no. 5, pp. 1113–1129, 2014.
5. M. K. Putchala, "Deep learning approach for intrusion detection system (ids) in the internet of things (iot) network using gated recurrent neural networks (gru)," Ph.D. dissertation, Wright State University, 2017.

Appendix A
A Survey on Malware Detection from Deep Learning

This appendix discusses a survey on malware detections from deep learning. Malware detections are also an important issue due to the increasing number of malware and similar approach as IDS.

A.1 Automatic Analysis of Malware Behavior Using Machine Learning

Recently, computer security faces the increase of security challenges. Because the static analysis is perceived as vulnerable to obfuscation and evasion attack, Rieck et al. [1] tries to develop dynamic malware analysis. The primary challenge in using dynamic malware analysis is the time needed to perform the analysis. Furthermore, as the amount and diversity of malware increases, the time required to generate detection patterns is also longer. Therefore, Rieck et al. propose malware detection method to improve the performance of malware detector based on behavior analysis. In this experiment, Malheur datasets were used. These datasets were created by themselves using behavior reports of malware binaries from anti-malware vendors, Sunbelt Software. Com. Each sample was executed and monitored using CW Sandbox's analysis environment and generates 3,131 behavior reports.

In this experiment, Rieck et al. [1] used four main steps. First, malware binaries were executed and monitored in a sandbox environment. It would give system calls and arguments as the output. Then, in step 2, the sequential reports produced from the previous step were embedded into a high-dimensional vector space based on its behavioral pattern. By doing this, the vectorial representation geometrically could be analyzed, to design clustering and classification method. In step 3, the machine-learning techniques were applied for clustering and classification to identify the class of malware. Finally, incremental analysis of malware's behavior was done

K. Kim et al., *Network Intrusion Detection using Deep Learning*,
SpringerBriefs on Cyber Security Systems and Networks,
https://doi.org/10.1007/978-981-13-1444-5

by alternating between clustering and classification step. The result shows that the proposed method successfully reduces the run time and memory requirement by processing the behavior reports in a chunk. Incremental analysis needed 25 min for processing the data, while regular clustering took 100 min. Furthermore, the regular clustering required 5 Gigabytes of memory during computation while incremental analysis only needed less than 300 Megabytes. So, it can be concluded that incremental technique in behavior-based analysis gives better performance in time and memory requirement than regular clustering.

A.2   Deep Learning for Classification of Malware System Call Sequences

Nowadays, the number and variety of malware are kept increasing. As a result, malware detection and classification need to be improved to do safety prevention. This paper wants to model malware system call sequences and uses it to do classification using deep learning. The primary purpose of leveraging machine learning in this experiment is to find a fundamental pattern in a large dataset. They used malware sample dataset gathered from Virus Share, Maltrieve, and private collections. There are three main contributions in this paper. First, Kolosnjaji et al. built DNN and implemented it to examine system call sequences. Then, in order to optimize malware classification process, convolution neural networks and RNN were combined. Finally, the performance of their proposed method was analyzed by examining the activation pattern of the neural unit.

During malware classification process, Kolosnjaji et al. [2] utilized malware collection from the dataset as the input of Cuckoo Sandbox. Then, the sandbox would give numerical feature vector as the output. After that, they used TensorFlow and Theano framework to construct and train the neural networks. It would give a list of malware families as the output of the NN. The neural network consisted of two parts, convolutional part and recurrent part. Convolutional part consisted of convolutional and pooling layer. At first, convolutional layer captured the correlation between neighboring input vectors and generated new features, resulting in feature vectors. Then, the result of the convolutional layer was forwarded to the input of recurrent layer. In recurrent layer, LSTM cells were used to model the resulting sequence and sort the importance based on mean pooling. Finally, dropout and a softmax layer were used to prevent overfitting to occur in the output. As the experiment result, it showed that the combination of convolutional network and LSTM gave better accuracy (89.4%) compared to feedforward network (79.8%) and convolutional network (89.2%).

A.3   Malware Detection with Deep Neural Network Using Process Behavior

The background problem of this paper is detecting whether there is malware infection on a computer based on the data traffic. To analyze this, usually, expert knowledge is needed, and the amount of time consumed is not a little. As a result, the purpose of this paper is to propose a method to detect malware infection using traffic data by utilizing machine learning. Tobiyama et al. leveraged recurrent neural network (RNN) for feature extraction and CNN for classification. This paper is good because, during the training phase using RNN, they used LSTM. RNN is known with the error vanishing problem because of its sequential structure. Its output depends on the previous input. As a result, when the previous input is getting bigger over time, an error will occur. LSTM avoids the error problem by selecting the only required information for future output to reduce the number of data.

In this paper, Tobiyama et al. [3] used 81 malware log files and 69 benign process log files for training and validation. The dataset was generated by using Cuckoo Sandbox to run malware files in an emulated environment. Then, they traced malware process behavior to determine generated and injected processes. The methodology of this paper is as follows: (1) generating log files during behavior process monitoring; (2) feature extraction using RNN, based on the log files from step 1; (3) converting the extracted features into image features; (4) training the CNN by using the image features; and (5) evaluating the process of validation using trained model. The result shows that the proposed model achieved 92% of detection accuracy. The drawback is the dataset was too small; they performed 5 min logging for 10 times, so the proposed system is not tested for large-scale data.

A.4   Efficient Dynamic Malware Analysis Based on Network Behavior Using Deep Learning

Nowadays, malware detection methods can be divided into three categories: static analysis, host behavior-based analysis, and network behavior analysis. Static analysis method can be evaded using packing techniques. Host behavior-based analysis can be deceived by using code injection. As a result, network behavior analysis becomes the spotlight because it does not have those vulnerabilities, and the need to communicate between the attacker and the infected host makes this method effective. One main challenge that hinders the use of network behavior analysis in malware detection is the analysis time. Various malware samples need to be collected and analyzed for a long period because the user does not know when the malwares start their activity. The main idea proposed by Shibahara et al. of this paper is aimed at two characteristics of malware communication, the change in communication purpose and common latent function. For the dataset, firstly they collected malware samples from VirusTotal, which were detected as malware by antivirus program. Then, they used malware samples in VirusTotal that have different sha1 hash compared to previously collected malware samples. They used 29,562 malware samples in total for training, validation, and classification.

Their methodology consists of three main steps: feature extraction, neural network construction, and also training and classification label. First, Shibahara et al. [4] extracted features from communications, which are collected with dynamic malware analysis. Then, these features are used as inputs of recurrent neural network (RNN). Then, in NN construction phase, the change in communication purpose was captured. During the training and classification, the feature vectors of root nodes were calculated according to VirusTotal. Then, based on the vectors, it gave the classification result. During the experiment, the analysis time and time reduction between the proposed method and the regular continuation method were compared. The result shows that the proposed method reduces 67.1% of analysis time and keeps the range of covered URL to 97.9% compared to full analysis method.

A.5   Automatic Malware Classification and New Malware Detection Using Machine Learning

The development of malware is going rapidly. Various malwares keep appearing, increasing the diversity of malware family. As a result, traditional static-based malware analysis cannot detect these new kinds of malwares. Because of that reason, Liu et al. [5] propose a machine-learning-based malware analysis system. In order to support this research, a significant amount of malware information was collected by using ESET NOD32, VX Heavens, and Threat Trace Security from their campus network. Approximately 21,740 malware samples that belong to 9 families, including viruses, worms, Trojan, backdoor, etc., were collected. A total of 19,740 samples were used for training, and 2,000 samples were used for testing.

The methodology consists of three main modules: data processing, decision-making, and malware detection. Data processing module includes gray-scale images, Opcode n-gram, and import function to extract the malware features. Decision-making module does the classification process and identifies the malware based on the features extracted from data processing module. It consists of Classifier1, Classifier 2, Classifier 3, until Classifier N. Finally, clustering process in malware detection module is used to discover new kind of malware by using Shared Nearest Neighbor (SNN) clustering algorithm. The result of clustering is either unambiguous or ambiguous. For the ambiguous one, SNN was implemented to identify it. In the experiment, the performance of the proposed method was tested by using several classifiers like random forest, K-Nearest Neighbor (kNN), gradient boosting, naive Bayes, logistic regression, SVM, and decision tree. The result of the experiment shows the average accuracy of the proposed method is 91.4%, while the best accuracy is 96.5% by using the random forest classifier.

A.6   DeepSign: Deep Learning for Automatic Malware Signature Generation and Classification

Recently, a variety of malwares keep growing, including a new variant of malwares that are undetected by antivirus software. Several methods have been proposed to overcome this problem, for instance, by using signatures based on specific vulnerabilities, payloads, and honeypots. However, all these methods have one big problem; they target a specific aspect of malware. As a result, if an adversary modifies small parts of their malware, it will not be detected by them. Because of that background, David et al. [6] proposes a method for a signature generation that does not depend on a particular part of malware, so it can be resistant to code modifier. In this research, David et al. used a dataset that consists of six categories of malware, such as Zeus, Carberp, Spy-Eye, Cidox, Andromeda, and DarkComet. They used a total of 1800 samples, which is 300 samples for each category.

The methodology consists of four main parts. First, malware program was run in Cuckoo Sandbox, an emulated environment. The output of the sandbox was sandbox log file. Then, the log file was converted into a binary bit string. The bit string was used as the input of Deep Belief Neural Network (DBN) to produce 30 sized vectors as its output layer. By doing that step, malware signature was generated. The question here is how they convert sandbox files to fixed size input. In order to do that, first, David et al. extracted all unigrams for each sandbox file in the dataset. Then for each unigram, the number of files that it appears was counted. After that, they selected top 20,000 with the highest frequency and finally converted each sandbox file to a 20,000 sized bit string. During the experiment, they used 1,200 samples for training (200 samples for each malware category) and 600 samples for testing (100 samples for each category). The number of features generated is 30, and the accuracy is 98.6%. The input noise was also set as 0.2 and learning rate as 0.001.

A.7   Selecting Features to Classify Malware

Since the development of machine learning several years ago, there have been various ideas to implement machine learning as an engine in malware detection software. However, because there is a time delay between malware landing on user's system and the signature generation process, it can harm the user. Because of that reason, Raman et al. utilized data mining to identify seven key features in Microsoft PE file format that can be used as input to the classifier. The seven features would be used in machine-learning algorithm to do malware classification. Raman et al. generate their dataset from PE files. Firstly, they write a parser to extract features from PE file. They use their experience in malware analysis to select a set of 100 features from the initial 645 features. Finally, they created a dataset of 5,193 dirty files and 3,722 clean files to evaluate those 100 features.

The focus of the methodology is feature extraction and feature selection, with the addition of combining intuitive method and machine-learning method during feature selection. First, Raman et al. [7] used their knowledge to reduce the number of the feature from 645 to 100. Then, random forest algorithm was utilized to choose 13 features. Finally, four classifiers (J48Graft, PART, IBk, and J48) were used to check the accuracy of each feature and choose the highest seven features. Those features are debug size that denotes the size of the debug directory table, image version that denotes the version of the file, debugRVA that denotes the relative virtual address of the import address table, ExportSize that denotes the size of the export table, ResourceSize that denotes the size of the resource section, VirtualSize2 that denotes the size of the second section, and NumberOfSections that denotes the number of section.

A.8 Analysis of Machine-Learning Techniques Used in Behavior-Based Malware Detection

The main problem in this paper is about the increase of malware varieties, which lead to vulnerable manual heuristic malware detection method. In order to cope with this problem, Firdausi et al. proposed an automatic behavior-based malware detection method utilizing machine learning. In this research, they used datasets in the format of Windows Portable Executable. The dataset consisted of 250 benign instances collected from System 32 of Windows XP 32 bit SP2. They also collected 220 malware samples from various resources. For monitoring process, Firdausi et al. [8] used Anubis, a free online automatic dynamic analysis service to monitor both malware and benign samples. Then, the performance of five classifiers was compared using their proposed method. The five classifiers were kNN, naive Bayes, J48 DT, SVM, and Multi-Layer Perceptron (MLP).

For the methodology of the research, firstly Firdausi et al. did data acquisition from community and virology.info. Then, the behavior of each malware was analyzed on an emulated sandbox environment. Anubis Sandbox was chosen for API hooking and system call monitoring. After that, the report would be processed into sparse vector model. The report was generated in xml format. For the next step, XML file parsing, feature selection, and feature model creation were done based on that XML file. Finally, the last step was doing classification based on that model. Firdausi et al. applied machine-learning tools, did parameter tuning, and finally tested their scheme. The result of this experiment shows that feature extraction reduces the attributes from 5,191 to 116 attributes. By performing feature selection, the time consumed to train and build the model became shorter. The overall best performance was achieved by J48 classifier with 94.2% true positive rate, 9.2% False Positive Rate (FPR), 89.0% precision, and 92.3% accuracy.

A.9   Malware Detection Using Machine-Learning-Based Analysis of Virtual Memory Access Patterns

For several years, people know that traditional malware detection can be divided into the static and dynamic method. These methods are usually implemented in antivirus software. Static method uses signature database to detect malware; on the other hand, dynamic method runs the suspicious program to check its behavior, whether it is a malware or not. However, there is a problem here. The software is vulnerable to malware exploit during infection, and it can be disabled by malware. So, in this paper, Xu et al. proposed an idea of hardware-assisted detection mechanism which is not vulnerable to such disabling problem. However, this idea relies on expert knowledge of the executable binary and its memory layouts. So, as a solution, they decided to use machine learning to detect a malicious action of malware. The primary purpose of this paper is to learn one model for each application that separates malware-infected execution from legitimate execution. Furthermore, Xu et al. will do the classification based on memory access pattern.

In order to monitor the memory access, Xu et al. [9] did epoch-based monitoring. A monitoring method divided program execution into epochs. Then, each epoch was separated by inserting a sign in a memory stream. They found that for most malicious behavior, the deciding feature was the location and frequency of memory accesses rather than their sequence. Then, after the monitoring had finished, the classifier was trained by using summary histograms for epochs. During training, the program was executed, and each histogram was labeled either malicious or benign. After the training model had been defined, the binary signature was verified, and the model was loaded into hardware classifier. Finally, the last step was hardware execution monitoring. If malware were detected, an authenticated handler would be launched automatically. During the experiment, they used three classifiers (SVM, random forest, and logistic regression) and compared their performance. The best performing classifier was random forest, with 99% true positive rate and less than 1% FPR.

A.10   Zero-Day Malware Detection

Lately, the variety of malware keeps increasing and threating the security of our computer system. People cannot rely on traditional signature-based antivirus anymore. Zero-day malware will easily bypass regular antivirus because their signature is not in the antivirus database yet. To solve this problem, Gandotra et al. [10] proposed a combination of static and dynamic malware analysis with machine-learning algorithm for malware detection and classification. However, there were several problems with this scheme. First, this scheme had high FP and False Negative Rate (FNR). Second, it took time to build the classification model because of the large dataset. As a result, early malware detection was not possible with this scheme.

Knowing these problems, Gandotra et al. concluded that the challenge here was to select the relevant set of features so the building time could be reduced and the accuracy would be improved. The dataset that is used in this experiment was taken from VirusShare. About 3,130 portable executable files, which include 1,720 malicious and 1,410 clean files, were utilized. All files were executed in a sandbox to get their attributes and then were used to build the classification model using WEKA.

The methodology consisted of six main steps. The first step was data acquisition. In this step, they collected malware samples targeting Windows OS from VirusShare database. They also collected clean files manually from system directories of Windows. The second step was automated malware analysis. In this phase, they used modified Cuckoo Sandbox to execute the specimen and generate the result as Java Script Object Notation (JSON) file. The third step was feature extraction. In this step, the JSON reports generated by Cuckoo Sandbox were parsed to obtain the various malware features. The result was a feature set of 18 malware attributes which can be used to build the classification model. The next step was feature selection. They selected seven top features by using IG method. IG method is an entropy-based method for feature evaluation which is broadly used in machine learning. The last step was classification. They used the selected seven features to build the classification model using machine-learning algorithm in WEKA library. The used seven classifiers are IB1, naive Bayes, J48, random forest, bagging, decision table, and multilayer perceptron. The result of their experiment showed that random forest gave the best accuracy with 99.97% and the time to build model was 0.09 s.

References

1. K. Rieck, P. Trinius, C. Willems, and T. Holz, "Automatic analysis of malware behavior using machine learning," *Journal of Computer Security*, vol. 19, no. 4, pp. 639–668, 2011.
2. B. Kolosnjaji, A. Zarras, G. Webster, and C. Eckert, "Deep learning for classification of malware system call sequences," in *Australasian Joint Conference on Artificial Intelligence*. Springer, 2016, pp. 137–149.
3. S. Tobiyama, Y. Yamaguchi, H. Shimada, T. Ikuse, and T. Yagi, "Malware detection with deep neural network using process behavior," in *Computer Software and Applications Conference (COMPSAC), 2016 IEEE 40th Annual*, vol. 2. IEEE, 2016, pp. 577–582.
4. T. Shibahara, T. Yagi, M. Akiyama, D. Chiba, and T. Yada, "Efficient dynamic malware analysis based on network behavior using deep learning," in *Global Communications Conference (GLOBECOM), 2016 IEEE*. IEEE, 2016, pp. 1–7.
5. L. Liu, B.-s. Wang, B. Yu, and Q.-x. Zhong, "Automatic malware classification and new malware detection using machine learning," *Frontiers of Information Technology & Electronic Engineering*, vol. 18, no. 9, pp. 1336–1347, 2017.
6. O. E. David and N. S. Netanyahu, "Deepsign: Deep learning for automatic malware signature generation and classification," in *Neural Networks (IJCNN), 2015 International Joint Conference on*. IEEE, 2015, pp. 1–8.
7. K. Raman et al., "Selecting features to classify malware," *InfoSec Southwest*, vol. 2012, 2012.

8. I. Firdausi, A. Erwin, A. S. Nugroho, *et al.*, "Analysis of machine learning techniques used in behavior-based malware detection," in *Advances in Computing, Control and Telecommunication Technologies (ACT), 2010 Second International Conference on*. IEEE, 2010, pp. 201–203.
9. Z. Xu, S. Ray, P. Subramanyan, and S. Malik, "Malware detection using machine learning based analysis of virtual memory access patterns," in *Proceedings of the Conference on Design, Automation & Test in Europe*. European Design and Automation Association, 2017, pp. 169–174.
10. E. Gandotra, D. Bansal, and S. Sofat, "Zero-day malware detection," in *Embedded Computing and System Design (ISED), 2016 Sixth International Symposium on*. IEEE, 2016, pp. 171–175.

Printed in the United States
By Bookmasters